100 THINGS

YOU DON'T NEED A MAN FOR!

Home Repair and Improvement

100 THINGS

YOU DON'T NEED A MAN FOR!

Alison Jenkins

AURUM PRESS

First published in Great Britain in 2001 by

AURUM PRESS LTD

25 Bedford Avenue

London WC1B 3AT

This book was conceived, designed, and produced by

THE IVY PRESS LIMITED

The Old Candlemakers, West Street

Lewes, East Sussex BN7 2NZ

Creative Director: PETER BRIDGEWATER

Publisher: SOPHIE COLLINS

Editorial Director: STEVE LUCK

Art Director: CLARE BARBER

Mac Design: GINNY ZEAL

Senior Project Editor: CAROLINE EARLE

Photographer: IAN PARSONS

Additional photography: CALVEY TAYLOR-HAW

Illustrators: ANNA HUNTER-DOWNING, IVAN HISSEY

Picture Researchers: LIZ EDDISON, VANESSA FLETCHER

A catalogue record for this book is available from the British Library

ISBN 1 85410 812 3

Originated and printed by Hong Kong Graphic and Printing Ltd, China

Contents

TOOL BELTS ARE SO THIS SEASON!

INTRODUCTION 6

TOOLING UP 8

FIXING IT 26

LAYING IT 50

NAILING IT 78

STORING IT 110

FAKING IT 126

NOT DOING IT 172

HOUSE RECORD 180

TEMPLATES 182

GLOSSARY 186

FURTHER READING AND USEFUL ADDRESSES 188

INDEX 190

ACKNOWLEDGEMENTS 192

DOES DONNA KARAN DO SAFETY GOGGLES?

SOME WORDS OF WISDOM

Introduction

MOST OF THE people I've spoken to about this book have either laughed out loud or allowed themselves a little chuckle at least – even the blokes! This book is not intended to be a dull manual that will send you off into a deep sleep within the first three pages, nor is it aimed at baffling the reader with technical terms and complicated diagrams. I'm just trying to give it to you plain and simple, with my tongue firmly in my cheek – the direct approach is always best.

My mother always used to say: 'If you want something done, then do it yourself.' I suppose this has been indelibly printed on my mind ever since I was a child. Some people are born practical, some achieve practicality and some have practicality thrust upon them. I'm hoping that this book will help you, whichever category you care to put (or find) yourself in. Although I do also have my dad to thank for nurturing in his daughter a healthy interest in wonderful gadgets and useful tools. He once gave me a house-warming present of two little crowbars and a rubber mallet, saying excitedly: 'Crowbars are always useful and you never know when you might want to hit something and not leave a mark.' How right he was – and I did like my gift, a lot.

Most of the basic DIY tasks are pretty simple; all you need to know is what to do, how to do it and what to do it with. If, like myself, you find yourself the proud owner of your first new home, and also the owner of a very undernourished bank account, then you will certainly need this book. It is highly unlikely that your new nest is perfect in every way and needs no remedial or cosmetic attention at all. So what are you going to do? Call some guy in to do the repairs for you and do

❋ GEE, THE WASHER'S BROKEN AND ALL MY CLOTHES ARE IN IT. WHO SHOULD I CALL?

READ *100 THINGS YOU DON'T NEED A MAN FOR!* AND THE PROBLEM'S SOLVED.

without all treats and new shoes for ever? Which would you prefer – paying for hefty call-out fees or a new handbag? No contest as far as I'm concerned, the new handbag wins every time!

In my opinion, there's nothing to stop a woman doing her own home repairs and decorating, and if a man wants to impress you there are lots of other ways to do it – that's why God invented champagne and restaurants, isn't it? But seriously, girls – read on. You will learn to love your toolbox, enjoy a trip around the local DIY store and you won't look at wood in the same way again. Yes, you can learn how to fix it, lay it, nail it, store it, fake it and then learn about not doing it too. This book is a journey through the basics of home repair, maintenance and decoration with lots of easy-to-grasp instructions and non-baffling information about everything from the front door to the kitchen cabinets! Also included are original and inexpensive home-improvement projects – the key on the right shows the level of ease, cost and the time it will take.

To conclude, here are some words of wisdom before you pick up that toolbox with confidence and become a DIY diva. Confucius said: 'In all things, success depends upon previous preparation, without which there is sure to be failure.' You don't need to chant this DIY mantra all the time, but bear in mind that preparation, as you will hear throughout this book, is truly the key to success.

So, girls, have fun trying out the 100 things you don't need a man for, then you can go out shopping and spend all the money you have saved! Enjoy!

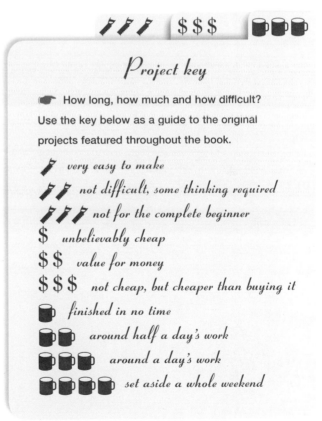

WE WON'T BE NEEDING ONE OF THESE ANY MORE – NOT FOR DIY ANYWAY.

Project key

☛ How long, how much and how difficult? Use the key below as a guide to the original projects featured throughout the book.

🪚 *very easy to make*

🪚🪚 *not difficult, some thinking required*

🪚🪚🪚 *not for the complete beginner*

$ *unbelievably cheap*

$ $ *value for money*

$ $ $ *not cheap, but cheaper than buying it*

☕ *finished in no time*

☕☕ *around half a day's work*

☕☕☕ *around a day's work*

☕☕☕☕ *set aside a whole weekend*

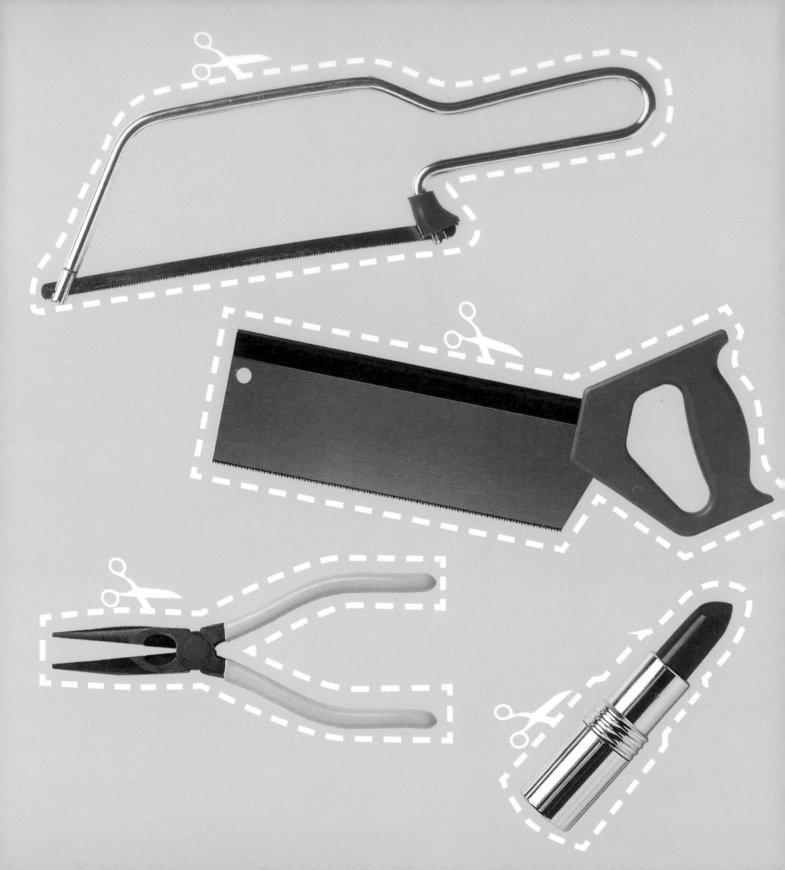

TOOLING UP

All a girl needs to fix it, lay it, nail it, store it and fake it. As easy as getting a make-up kit together!

Shopping

TOOL KIT! Now, those words may not excite you as much as the words 'new shoes' would, but believe me, a girl needs to have the right tool to match the job (just as you need the right shoes to match the outfit). While I do appreciate that most of you will have much better things to spend your hard-earned cash on than an extensive new tool kit, a few basic items won't break the bank and I assure you that in the fullness of time they will prove to be invaluable.

HAND TOOLS

In general, tools fall into two categories: 'hand' and 'power'. Hand tools are the ones that don't use any electrical power whatsoever, either from the mains or batteries. In other words, you have to do all the work yourself. You'll see here a small selection of essential ingredients to get your tool kit started.

Flat-head screwdriver

Cross-head screwdriver

Spirit level

An essential tool if you don't want to live in a crooked world. They come in all sizes, but a medium one will do as long as it has two bubbles so you can see the true horizontal and vertical (some even have a third bubble for 45-degree angles). Use it when putting up shelves, hanging wallpaper and so on.

Always check that a tool has a comfortable, firm grip before you buy.

Flat-nosed pliers

Long-nosed pliers

Screwdrivers

There are numerous sizes and shapes, but four will get you just about everywhere you need to go. The basic types are flat-head screwdrivers for screws with a slotted head, and cross-head (Posidriv or Phillips) for screws with a cross-shaped indentation in the head. Choose ones with nicely shaped handles that feel comfortable in your hand – a small and medium size of each type should suffice. You can also buy a multipurpose tool with one handle and interchangeable heads. Perfect!

Pliers

Set of three: long-nosed, flat-nosed and side-cutter. Used for general gripping and manipulation of metal, wire or anything narrow or circular. Side-cutters are good for snipping off ends of wire.

NOT JUST A HAMMER, IT'S A SHOPPING OPPORTUNITY!

Claw hammer

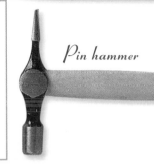

Pin hammer

SHARP STUFF

Craft knife Most of you will have one of these – the type with snap-off blades – for cutting paper and other light materials.

Stanley knife A heavier-duty version of the craft knife; it takes double-ended replaceable blades. Use it for cutting and scoring tougher materials such as thick vinyl or carpeting.

Stanley knife

Hammers

Two will do for starters: a small pin hammer for use with panel pins and small nails, and a heavier claw hammer for banging in bigger things and removing old nails Choose ones that have a comfortable grip and are not too heavy to use.

SAWS

Tenon saw A rigid saw for cutting thicker bits of wood for simple woodwork.

Junior hacksaw A small saw with a replaceable blade for cutting small-bore metal pipes or thin wood.

Tenon saw

Chisels

A starter set of three (small, medium and large) is a useful addition to the tool kit if you intend to do any woodwork. Use a chisel to cut recesses for locks and hinges and for making simple joints.

Chisel

CAN'T DO WITHOUT

Bradawl A pointed metal spike set into a handle, used for making a pilot hole before inserting a screw into wood (there is a danger of splitting the wood if you forget to make the pilot hole first). Also very useful for all sorts of hole-making.

Try square This little gadget makes life easier when cutting wood that has to be perfectly square. The steel blade is fixed at 90 degrees to the handle. Hold the handle flush with the edge of the wood, then mark your cutting line with a pencil along the blade.

Adjustable wrench/spanner A pretty good tool to have; the adjustable jaw makes it a one-size-fits-all type of gadget.

Gripping pliers Looks like a mini mechanical dinosaur, with gripping jaws that can be adjusted and locked to suit the task. A kind of plier-cramp hybrid.

Retractable steel tape measure Choose one that is about 5 m (16 ft) long and can be locked at any point. This enables you to measure long or large objects easily.

Straightedge Handy for cutting straight edges when you are using a Stanley knife or drawing straight edges. Take great care though, DIY depends on having a prehensile thumb.

Just plug in and go!

POWER TOOLS

Yes! You've got the power! Power tools are fantastic. Once you have experienced the power, you will never look back. Power tools make DIY life easier, quicker and altogether more enjoyable. Cordless tools are even more desirable and wonderful. As the name suggests, they have no flex and are powered by an internal rechargeable battery, so you can wander about wherever you like!

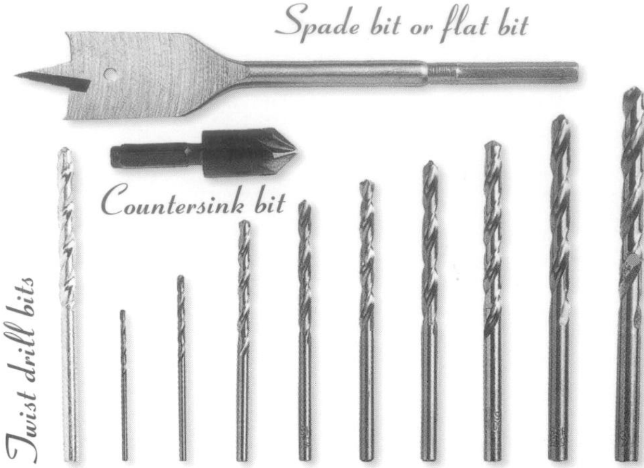

✻ ALL TOOLED UP AND DRESSED TO DRILL!

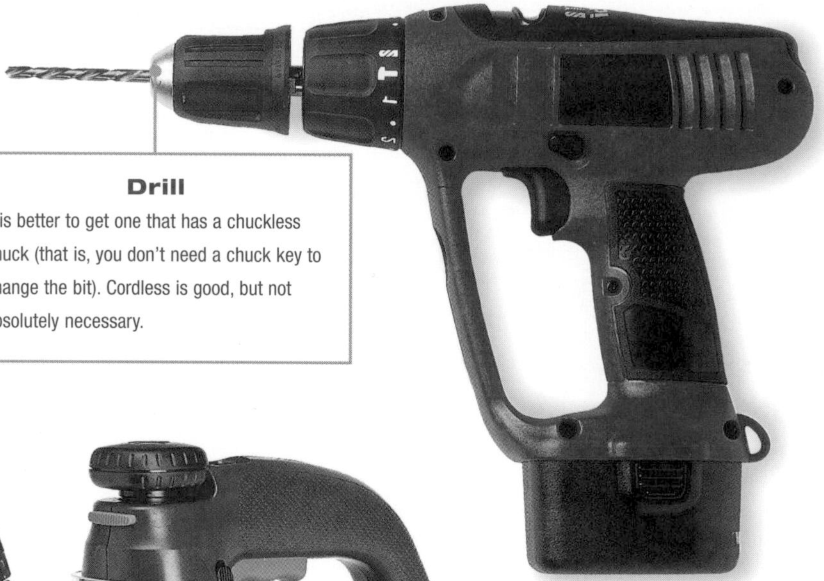

Spade bit or flat bit

Countersink bit

Twist drill bits

Drill bits

Get a set of all-purpose drill bits as a starting point and then add to the collection as necessary. A few flat-head bits are useful for cutting larger holes, and a countersink bit makes a little recess for the screwhead to nestle in neatly.

Drill

It is better to get one that has a chuckless chuck (that is, you don't need a chuck key to change the bit). Cordless is good, but not absolutely necessary.

AND THEY COME IN DIFFERENT COLOURS TOO...

Jigsaw

A jigsaw changed my life and will change yours too. In general, a jigsaw is used for cutting curves in wood, boards and even metal or ceramic with the correct blade. You can cut any angle or shape, and if you want to get really artistic, buy a scrolling jigsaw with an adjustable knob for intricate shapes.

12

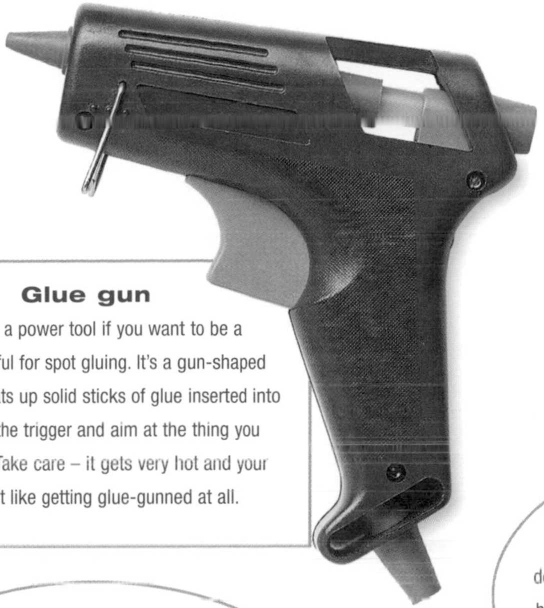

Orbital sander

Glue gun

Not technically a power tool if you want to be a purist, but useful for spot gluing. It's a gun-shaped object that heats up solid sticks of glue inserted into the back. Pull the trigger and aim at the thing you need to stick. Take care – it gets very hot and your fingertips won't like getting glue-gunned at all.

Power sander

Another labour-, blood-, sweat- and tears-saving device. Sanding anything, unless it is a very small job, is hard work. A detail sander has a shaped sanding pad to enable you to get into small or awkward corners. Sanding is still hard work even with a power sander, but less so.

Power screwdriver

You may think that this is unnecessary, but try constructing flatpack furniture without one. It takes the sweat out of screwing and unscrewing (it also has a reverse function).

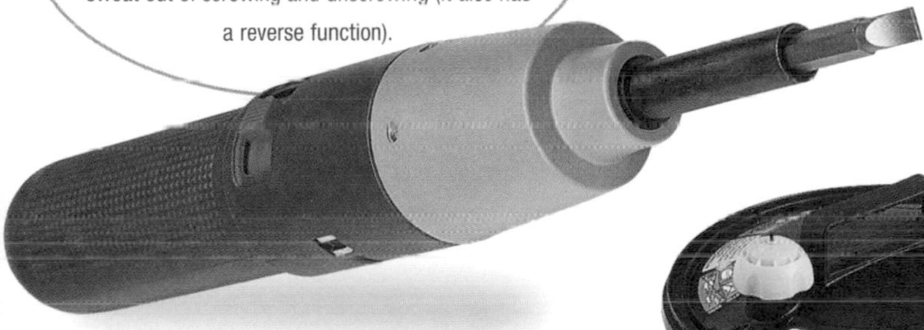

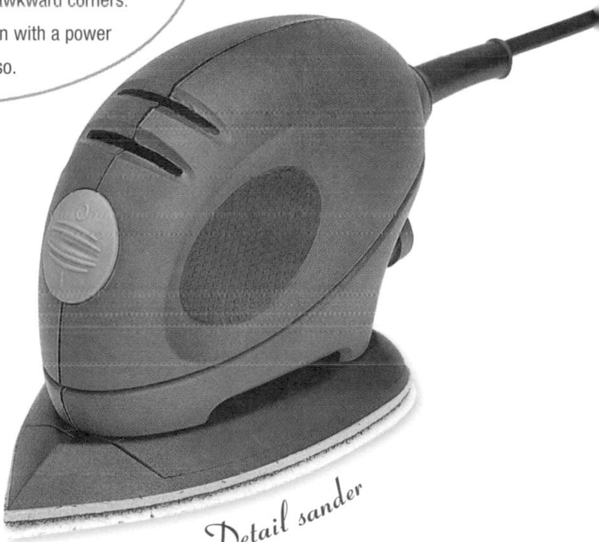

Detail sander

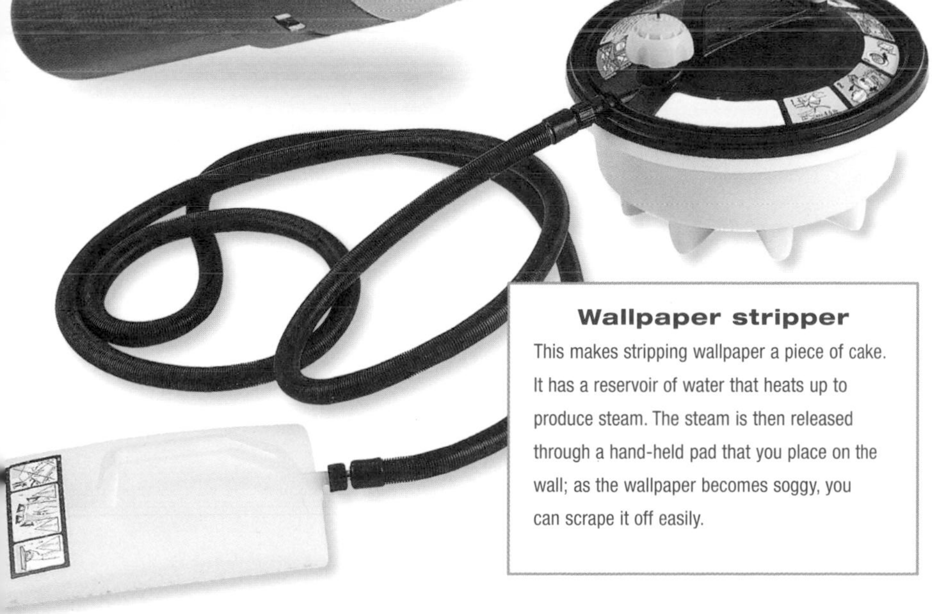

Wallpaper stripper

This makes stripping wallpaper a piece of cake. It has a reservoir of water that heats up to produce steam. The steam is then released through a hand-held pad that you place on the wall; as the wallpaper becomes soggy, you can scrape it off easily.

AND CAN'T DO WITHOUT

Hot air gun Although it looks like a hair drier, this handy gadget emits a fierce heat that melts paint so that you can remove it with a scraper or shave hook. Just pull the trigger and point it at the paint to be stripped!

PAINTING AND DECORATING EQUIPMENT

Self-explanatory really – all the stuff you'll need for decorating your home. Stuff for getting it off, getting it on and hanging it up. It's worth investing in these basics to make decorating life a little easier. Substitutes just don't do the job – you can't scrape wallpaper off with anything else but a wallpaper scraper.

JUST LIKE MAKE-UP – PICK THE COLOURS YOU LIKE AND APPLY WITH A BRUSH!

GETTING IT ON

Paintbrushes A huge variety of paintbrushes can be purchased for any sort of painting task your heart desires. Start with a basic set of three: 100 mm, 50 mm and 25 mm (4 in, 2 in and 1 in) for general painting. A chisel-edge brush with angled bristles is very useful for painting around windows, doors or fittings. Other brushes can be bought for various paint effects, such as colour washing or dragging.

Paint pad An alternative to brushes or rollers. Easy and quick to use on flat surfaces, economical on paint too. Use in conjunction with a paint tray.

Basic rollers Paint rollers come in 180 mm (7 in) and 230 mm (9 in) standard sizes. The removable sleeve can be made of foam, long pile or short pile. Special rollers have long handles for painting behind radiators or extension handles for painting ceilings.

Roller tray A plastic tray made to fit a paint roller. This has a recess to hold the paint and a flat rolling area to ensure that the roller sleeve is evenly coated.

Sponges Natural sponges can be used for a multitude of decorative paint effects.

Paintbrush

Paint kettle

Little metal or plastic containers with handles, used for decanting or mixing paint in.

Paint rollers

Paintbrushes

String

Tie a piece of string between the legs of the wallpapering table at one end. Tuck the end of the paper behind the string to stop it from curling up when you try to cut or paste it.

It's important that your wallpaper knows its place.

Dragging brush

Filling knife

Similar to the scraper but with a narrower blade, used for applying filler to holes and cracks for surface preparation.

Scraper

A knife with a broad flexible blade, used for removing old paint and wallpaper.

Plumb line

An essential item for hanging the strips of wallpaper on the true vertical. Yes, you may find that your walls aren't quite true, especially in older buildings. It's important to get the first strip hanging straight.

Shave hook

Tool with a shaped blade set at right angles to the handle for scraping paint from mouldings or awkward corners.

Seam roller

When you have hung all your wallpaper, you may find that the edges start to come unstuck. Just roll over with the seam roller to re-adhere.

Wallpapering scissors

These have extra-long blades for cutting and trimming wallpaper.

OTHER THINGS YOU NEED

Cleaners Most water-based paints are cleaned up with detergent and water, but oil-based or metallic paints need special cleaner or white spirit.

Cloths Essential for cleaning up those inevitable spills and also used for a few decorative effects.

Decorator's filler A general-purpose filler for holes, cracks and so on.

Wallpapering table Also essential. Folding tables are useful. Must be long enough to hold your wallpaper strips with a bit hanging over at each end.

Masking tape Use this self-adhesive tape to mask off anything you don't want painted, such as panes of glass or other paintwork. Remove when the job is done.

Dust sheets Huge pieces of lightweight cloth or plastic that you can fling over furniture, floors or anything you want to protect from dust and paint splashes.

LOOK LIKE A PRO...

Bucket Mix up your wallpaper paste in this.

Wallpaper paste For sticking non-pre-pasted wallpaper to the wall.

Wallpaper brush Big floppy brush used for pasting wallpaper.

Not just for the professionals

SPECIALIST BUT USEFUL

These are the kind of tools and nifty little gadgets that I get really excited about; they're not even the essential ones that you'll use all the time, but when that tricky DIY opportunity knocks – boy, you'll wonder what you did without them! Just try marking out a shelf to fit a less-than-square alcove without a bevel, or stripping wires neatly using a craft knife. Believe me, the addition of a few specialist items to your kit will undoubtedly save you time and energy.

WE RECOMMEND A RUBBER MALLET FOR THOSE TIMES WHEN YOU WANT TO HIT SOMETHING WITHOUT LEAVING A MARK!

Sliding bevel

A fantastic little tool used to find internal angles when, for example, you need to cut a shelf to fit an alcove that isn't perfectly square. Simply place the sliding bevel in the corner, open the blade and there's your angle. Use it to mark the correct angle on the material to be cut.

✱ SO MANY INTERESTING TOOLS AND SO LITTLE TIME.

Rubber mallet

Heavyweight hammer with rubber head, useful for hammering dowel joints home.

Portable workbench

I love my workbench. It folds up and is fully adjustable for height and all sorts of things. Not absolutely essential, but handy for when you are sawing something – better than using the kitchen table.

Glue spreader

Plastic knife with flexible blade, used to spread glue evenly and thinly.

Wire stripper

Another beast with sharp jaws. This ingenious little tool strips the insulation from electric wire without damaging the wires inside.

Surform planer

I recently purchased one of these, and I really do wonder how I did without it. Great for taking rough edges off wood and for reducing a piece to size – fixing sticky doors, rounding corners and so on. Blades are replaceable and come in a variety of sizes.

Small crowbars

These levers take the strain out of removing skirtings, panelling, plasterboard, large nails and floorboards.

Stud finder

Device for locating studs or joists through plasterboard or plaster. Attach screws at those locations for greatest strength.

Mitre saw

Yes, yet another tool that has changed my life. A mitre block is fine, but one of these is indispensable. Cut any angle you like, make perfectly straight cuts and make really neat angled joins.

You may not need a man, but you need a stud finder – it isn't what you think it is.

More stuff to make doing it yourself easier

MISCELLANEOUS EQUIPMENT USEFUL

Again, not absolutely essential, except perhaps the ladder, but useful. This will become clear when you try to undo a nut with a pair of pliers instead of a spanner, or try to hold something steady while you cut it without having a handy cramp, or indeed try wallpapering while teetering on a kitchen chair! Always get the right tools for the job.

Staple gun

This should possibly be with the essential tools, as a staple gun is a great tool for fixing loads of things, especially soft furnishings, making blinds and so on.

These come in different sizes

G-clamp

This looks like the letter 'G' in profile, hence the name, and is used to hold things still while you saw or glue.

* I'VE GOT MORE TOOLS AT MY DISPOSAL THAN MY WINNING SMILE.

Step ladder

Unless you are a very tall girl, you will need some access equipment. A lightweight folding step ladder should serve your basic DIY requirements. Many have a handrail at the top, and most have a flat shelf to hold paint trays and other tools while you work.

18

Pencils

Keep a few pencils in your toolbox for general marking. An HB grade is best – markings will be accurate and sharp but not too dark.

Adjustable spanners

Available in different sizes, each with an adjustable jaw. Not as strong as an open-ended spanner or a ring spanner, but the movable jaw enables you to deal with a large or worn nut or one coated with paint.

Adjustable wrench

Half plier, half spanner, with adjustable locking jaw. Squeeze the handles so the jaws grip the work, then tighten the adjusting screw so jaws snap together. Pull the release lever to loosen wrench.

Set of allen keys

Lots of flatpack furniture arrives with allen keys for use in putting it together. Keep a set handy in case you need to do some deconstruction or adjustment.

Socket set

For dealing with nuts and bolts, this set contains a ratchet handle for sockets and a handle with interchangeable cross-head and slot-head bits for screws.

HEY CAN'T DO WITHOUT

Scissors Keep a pair of all-purpose scissors handy for cutting and trimming. Don't use dressmaking scissors to cut paper – it will ruin the blade.

Compass You may have one of these left over from your old geometry set. Strike an arc or draw a circle.

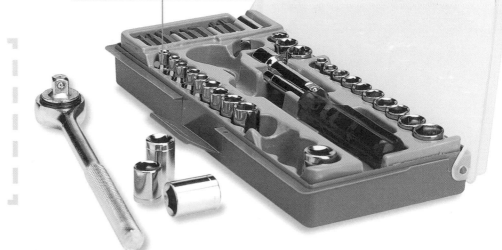

Getting to grips with the hard stuff

HARDWARE

Here is an assortment of common fixings: to hold things up, keep things together and fix small things to big things. There are different types of fixing, each designed to perform a particular task. DIY success depends on choosing the right one. You'll find useful information printed on the packets to help you make the right choice.

Escutcheon nail

Lath nail

Hardboard panel pins

Lost-head nail

Plain-head wire nail

Panel pin

Masonry nail

One-pin picture hook

Picture rail hook

Two-pin picture hook

❋ MMM, DON'T YOU JUST LOVE THESE ACCESSORIES!

NAILS

☞ Nails are made of metal and come in all shapes and sizes. Each is designed for a different purpose or for use with a particular kind of material. Panel pins are fine nails that vary in size from 15 mm (⅝ in) to 4 cm (1½ in) and are used for fixing panels or moulding to wooden frameworks or for making small joins. Escutcheon nails are used for attaching keyhole plates and other decorative backplates. If you're fixing thicker mouldings such as architraves, use lost-head nails; these have a head that can be hammered below the surface with a pin punch. Wire nails are used where they won't be seen, because the head remains visible. Masonry nails are the big tough ones used for attaching wood to masonry, brick or concrete. Other types of nails include floor brads for fixing floorboards to joists, glazing sprigs to hold panes of glass in the frame and upholstery nails and small tacks for general upholstery work.

Picture hangers

Traditional brass picture hooks are available in one- or two-pin form. Plastic picture hooks have a few small metal pins at the back that grip firmly when hammered into the wall. Picture rail hooks just hang from the picture rail. When hanging a heavy picture or mirror, use flat mirror plates. These brass plates are fixed to the back of the frame and can then be screwed securely to the wall.

Flat mirror plates

SCREWS

☛ In general, screws fall into two categories: slot-head, the ones with a straight indentation in the head, and cross-head, the ones with a cross-shaped indentation in the head (sometimes called Posidriv or Phillips). They can be raised-head, which sit above the surface and can also be decorative, or countersunk, when the head of the screw sinks below the surface into the countersink hole you have drilled previously. For woodwork, countersunk screws are the best. Like nails, screws come in a myriad of different diameters and lengths, each designed for a specific purpose. And the good thing about screws is that you can easily unscrew them.

Slot-head wood screws

Hardened steel slot-head screws

Cross-head screws

Plasterboard screws

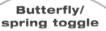

Butterfly/ spring toggle

Makes a very strong fixing in a cavity wall. The 'butterfly' springs open inside the cavity. When the screw is tightened, it is pulled towards the wall.

AND **Screw cups**

A little metal collar that slips onto the screw before it's inserted. The screwhead sits in the collar to make a neat, decorative fixing and also means that you do not need to drill a countersink hole first.

Plastic wall plugs

PLASTIC WALL PLUGS ENSURE A FIRMER GRIP!

VITAL INGREDIENT

Wall plugs When you want to fix something to a wall, you will need to drill a hole first, then insert a wall plug. The screw won't stay put in any type of masonry without one. Wall plugs are plastic reinforcements that line the drilled hole and create a good grip for the screw threads; this prevents the screw from pulling out under stress. When you buy wall plugs, the package label will show which size drill bit you will need. Simply drill the hole, insert the plug and tap it in with a hammer.

There are different types of plugs for solid walls and cavity walls of plasterboard. Solid walls are straightforward enough, but plasterboard is different because there's nothing behind it for the screw to hold on to. Hollow-wall plugs have a collapsible shaft and work on the principle that when the screw is driven in, the shaft opens up to grip the screw and hold it tight against the inside of the plasterboard.

Moving into the bathroom

SPECIAL TILING EQUIPMENT

Tiling seems like an impossible task, but the right tools make life so much easier. Choosing the tiles is not the only fun part – useful gadgets such as tile cutters, tile saws, tile nibblers and tile nippers will help you tile any kind of wall, no matter how many pipes and other objects get in the way.

Tile nipper and nibbler

Tile nippers are great for cutting tiles, especially mosaics, into specific shapes. Simply hold the mosaic tile in one hand and squeeze the jaws of the nipper on the edge. The tile will break cleanly in two. Try cutting tiles in half first, then experiment with smaller or irregular shapes. Practice makes perfect, so make sure you have a few spare tiles. The tile nibbler is similar to the nipper but has a small cutting area between the jaws. You can literally 'nibble' away at the edge of a ceramic tile to make small curved shapes to fit around pipes, etc.

Fancy a nibble before bedtime or are you too tired tonight?

Filling knife

You've seen this tool before. It can also be used for applying grout.

Bolster chisel

This has a wider blade than a regular chisel. Use it with a hammer for removing floor tiles. For wall tiles use a cold chisel (*see box opposite*). A bolster chisel is also useful for levering up floorboards.

Tile cutter

This is a hand tool that is used to score and cut tiles in perfectly straight lines. Simply place the tile in position on the cutter and then draw the blade firmly – just once – along the surface. Then place the tile in the cutting jaws of the tool and press the handle to make a clean cut. This takes a little practice, so make sure that you have a few spare tiles to hand.

Trowel

Scoop up adhesive with this tool, then spread it onto the wall. It has a pointed tip for getting in between tiles and into corners.

Tile spacers

Essential for evenly spaced tiles. Spacers are plastic cross-shaped objects that sit between the tiles, keeping each one the correct distance from its neighbour. And you can grout straight over them.

GROOVY

LOOK FOR INSPIRATION IN ALL THOSE STYLE MAGAZINES.

MORE ESSENTIALS

Cold chisel Use this with a hammer for chipping out broken tiles or for removing tiles completely.

Tile scorer/grout raker This handy tool has a very rough abrasive blade, and is tough enough to score tiles and remove old grout easily.

Notched spreader A plastic tool with notches so that the adhesive spreads evenly.

Grout shaper This tool looks a bit like a pencil and you use it to make a neat groove in the grout between tiles – very professional.

Tile saw Looks a bit like a hacksaw, but has a coarse round blade that makes it easy to cut curved shapes in ceramic tiles.

Tiles

For tiling you need tiles! Loads to choose: standard sizes and shapes, smaller tiles to fill in spaces, decorative mosaic tiles, borders and edge tiles, plus three-dimensional insets and feature tiles.

MISCELLANEOUS MATERIALS

BASIC

Along with the equipment already shown, here are other things you might need for masking, filling in, rubbing down and protecting yourself, plus a few things for emergencies. You don't need to get everything all at once – just as and when the task dictates – but you'll soon wonder how you ever coped without your trusty toolbox.

IF YOU LIKE IT ROUGH!!

❋ ALWAYS DRESS FOR THE JOB IN HAND... I'M OFF FOR DRINKS AND DINNER!

Sandpapers

Sandpaper is available in various grades from fine to coarse. You can buy assorted packs to see you through most woodworking tasks. Use wet and dry sandpaper for sanding metal. Flexible sanding pads are useful for sanding shaped surfaces. Rinse the pad with water when it becomes clogged. When sanding flat surfaces, wrap a piece of sandpaper around a cork sanding block.

BASIC SAFETY ITEMS

Unfortunately designed for safety, not style or comfort

Protective gloves Use when handling anything that is caustic, sharp, abrasive or indelible.

Dust mask It is vital to wear this when doing anything that involves dust, such as sanding, and when spray painting to prevent inhaling something dangerous or carcinogenic. Check the packet: some masks are designed for use in general dusty conditions and others are for more dangerous materials such as MDF (medium-density fibreboard).

Safety goggles Although not particularly comfortable, and certainly not fashionable, safety goggles are an essential basic safety item for certain DIY tasks. Always wear safety goggles when cutting or sawing most materials, particularly ceramic or mirror tiles and metal objects. The safety goggles will prevent you from getting any potentially blinding shards in your eyes as well as protecting your eyes from irritating dust.

Fine filler

Decorator's filler

Wood filler

Fillers

Choose a suitable filler for the surface, whether it is fine filler or decorator's filler for walls or wood filler for timber. Wood filler is available in different colours to match the type of wood to be filled.

Fuses

Spares

Keep all your spares in a safe, accessible place so you can get at them easily when the need arises: spare fuses, light bulbs, batteries and so on.

TOOLBOX

☞ The toolbox is not absolutely necessary, but in my opinion it is far better to have all your bits and pieces in one place – it saves time and energy when a DIY project presents itself. I have a metallic red one, which opens up to reveal compartments in which to store large and small items neatly. Many years ago when I began my collection, my toolbox was a rather bare specimen – today of course it is overflowing with all sorts of useful equipment. However, the bigger the toolbox, the heavier it is likely to be – your handy toolbox is not so handy if it requires you and a friend to lift it. I suggest that you keep the essentials in a portable box, and other larger, heavier or specialized items somewhere else.

Make-up

For essential retouches after a hard DIY session.

Tapes

Use masking tape when you need to protect woodwork or window panes when painting. Heavy-duty double-sided tapes are available for sticking a carpet to the floor and so on.

THINK OF IT AS A BIG HANDBAG MADE OF METAL!

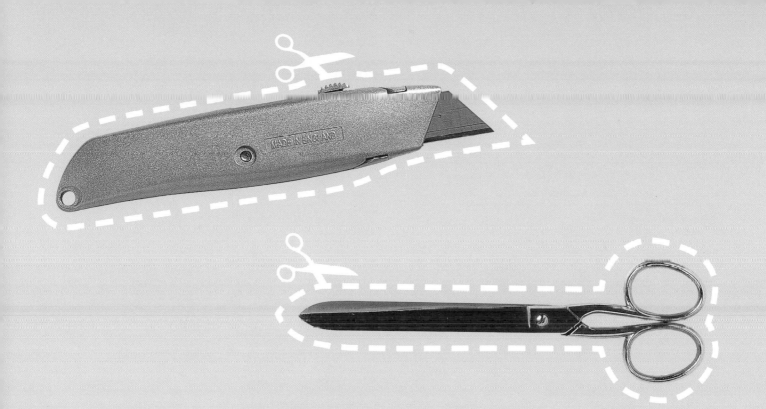

FIXING IT

All the odd jobs that need doing around the house, from changing a plug to unblocking a drain.

Odd jobs

NOT REALLY jobs that are odd, but just lots of those little annoying things that slowly but surely become hugely irritating. Most odd jobs are easily and quickly carried out – all it takes is a little motivation and know-how. When you've tackled the problem and the fault is fixed (which will probably take a fraction of the time spent complaining about it), pat yourself on the back proudly and wonder why you didn't do it ages ago! Whether it's changing a plug, stopping a draught, changing a lock or taking the squeak out of a floorboard or a door hinge, you'll find lots of information here to help you solve your odd-job problems. And you'll be amazed at how straightforward most of the odd jobs really are.

HOW NOT TO MAKE SPARKS FLY
ZAP!

Electricity is one of those things that is best left to the professionals. But changing a plug is pretty simple and it's probably better not to call the electrician to change a plug for you. Remember that electricity can be very dangerous, so it is essential that you take every precaution to protect yourself and others when dealing with any DIY task that involves electricity (*see box opposite*).

✽ IF YOU ARE IN ANY DOUBT ABOUT AN ELECTRICAL REPAIR, CALL A PROFESSIONAL!

WIRING DIAGRAMS

Electrical currents and wiring colours vary around the world, so depending on where you live your plug may look different. You may have to do a bit of visualization to relate the diagrams below to the plug you have in your hand. Nonetheless, when it comes to plug changing, the steps on the page opposite are pretty universal.

UK clamp terminal plug

✽ **EARTH** – green and yellow
✽ **NEUTRAL** – blue
✽ **LIVE** – brown

European clamp terminal plug

✽ **EARTH** – green and yellow
✽ **NEUTRAL** – blue
✽ **LIVE** – brown

US grounded plug

✽ **EARTH** – green or bare copper wire
✽ **NEUTRAL** – white
✽ **LIVE** – black

CHANGING A PLUG

Most new appliances have plugs attached, but you may need to change an old or broken plug, or perhaps over time you've been a bit rough so the cable has frayed or the wires pulled out of the plug. When your finances won't allow you to fling out the old and replace with brand new, it's time to repair

1 First you need to remove the screw that holds the old plug together and separate the two halves. Now, take a look inside and see if you can spot any similarities with the diagrams opposite – contrast and compare. Spotted one similar to yours? Good! If the plug is integral with the cable, just cut the cable close to the plug, then dispose of the old plug safely.

need
* wire stripper
* new plug
* Stanley knife
* small screwdrivers: cross-head and flat-head

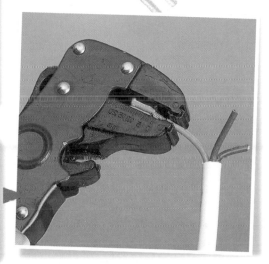

2 Using the wire stripper, carefully remove about 5 cm (2 in) of the thick outer sheath and separate the wires. The new plug usually comes with a diagram to show you how long to cut each wire. If your plug requires a fuse, check that it has the correct one: a 3-amp (red) fuse for appliances that are rated below 700 watts and a 13-amp (brown) fuse for those rated between 720 and 3000 watts. The earth wire is usually longer than the neutral and live wires to make it easier to fit them into the terminals.

3 Strip about 12–15 mm (½–⅝ in) from each of the three inner wires (conductors). Twist the filaments of each, using your thumb and forefinger, to make the ends neat. Loosen the large screw to remove the cover of the new plug. Place the flex on the open plug; our example has a sprung cord grip to prevent the flex from being pulled from the plug.

SAFETY PRECAUTIONS

☛ Turn off the power at the consumer unit before inspecting any switches, sockets or installation. Double-check that the power is off by plugging in an appliance.

☛ Always unplug an appliance first before attempting any repair.

☛ Check plugs and flexes regularly and repair or change worn flexes or broken or damaged plugs. Ensure that flexes are kept away from heat sources such as ovens or fires.

☛ Never use a fuse that is rated too high for the appliance.

4 Most plugs have post terminals, where the wire is held securely in the hole at the top by a small clamping screw. You need to fold the end of the twisted filaments back on itself, loosen the clamping screw, then insert the wire. Tighten the clamping screw; then tug gently to make sure it is held securely. Other types of plugs have clamp terminals; simply wrap the twisted wire around the post clockwise, then screw on the clamp to secure.

Problem doors and how to fix them

FIXING A ☞ STICKING DOOR

You may find that a wooden door in your home occasionally sticks or 'binds' in the door frame. This is more often than not caused by climatic changes such as dampness in the atmosphere. The wood will absorb excess moisture and possibly swell, causing the problem. Also, if you have changed your flooring material (switching a new thick carpet for the old threadbare one), there may be a need to shave a little from the bottom of the door so it fits neatly.

1 First take a look at the door in order to assess where the problem is. If it needs planing at the side, use a surform planer to shave the part of door that sticks, to ensure a proper fit. Revarnish or repaint the exposed wood. If the door needs planing at the top or bottom edge, you will have to remove the door to do so, then rehang it (*see pages 96–97*).

2 Hold the door firmly between your legs so that it remains still while you are planing.

* MMM, STRONG AND SILENT, JUST MY TYPE OF DOOR!

DO YOU HAVE AN EMBARRASSING PROBLEM WITH WIND?

FIXING A SQUEAKY DOOR

Initially you may not worry about a squeaky door, but after a while the noise can become infuriating. The squeak is usually caused by the hinges suffering from lack of lubrication. Simple remedy: squirt a few drops of oil on each hinge. General all-purpose light oil will do – the type that you get for bicycles – and the nicely lubricated hinge will fall silent.

DRAUGHTY LETTERBOX

If your letterbox tends to flap in the wind and let in more than the morning mail, you can fix this by attaching a letterbox excluder to the inside of the door to completely cover the opening. This is a rectangular plastic or metal frame that surrounds the letterbox and has bristles at the centre so that the postman can do his job while the cold winter winds are kept out.

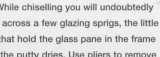

3 While chiselling you will undoubtedly come across a few glazing sprigs, the little nails that hold the glass pane in the frame while the putty dries. Use pliers to remove each sprig when you see it; don't try to knock them out with the chisel.

need

* glass cut to size
* safety gloves and goggles
* hammer or wooden mallet
* chisel and pliers
* putty and putty knife
* glazing sprigs

4 Take some putty in your hand (you can take the safety gloves off for this) and knead it into a thin sausage shape. Firmly squeeze the putty sausage into the recess all the way around the window frame.

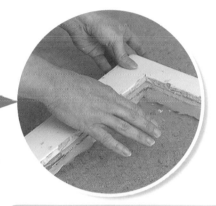

5 Insert the new pane of glass carefully, pressing against the putty all around the window frame. Then gently hammer in glazing sprigs 25 cm (10 in) apart all around the frame. Avoid hitting the pane too hard.

6 Roll some more thin putty sausages and press them against the edge of the glass all around the window frame. This should also cover the glazing sprigs. Use a putty knife to scrape off the excess and create a smooth finish inside and out.

SAFETY PRECAUTIONS

Glass can be an unpredictable thing; it can crack or shatter if accidentally knocked over or tapped with a hammer. Make sure that you put the new pane in a safe place while you complete the preparation work.

☞ Always wear thick safety gloves and safety goggles when removing or handling broken shards of glass.

☞ Make sure that you wear sturdy footwear when dealing with panes of glass, just in case you drop it.

☞ When buying your replacement pane, always tell the glazier what you need if for. He will make sure he gives you the right type and thickness of glass.

Keeping your home safe and secure

INSTALLING A SMOKE ALARM 6

The risk of fire in your home can easily be avoided by taking the proper safety precautions and by fitting a smoke detector. These potential life-savers come in the form of battery-operated, self-contained units and their function is to detect the presence of smoke or toxic fumes. The smoke alarm will issue an ear-piercing warning noise to alert you to the danger of a fire, enabling you to leave the house in good time.

WHERE IS THE BEST PLACE TO PUT IT?

Well, not in the kitchen or bathroom, or near a heater or air vent, as heat and steam can trigger an alarm. Ideally it should be situated on the ceiling, at least 30 cm (1 ft) away from any light fittings. If your home is on one floor, install the detector in the hallway between bedrooms and the living area. If you have two floors, put one above the foot of the stairs and another on the upstairs landing.

Remember that heavy-duty sanding can set off your smoke alarm too!

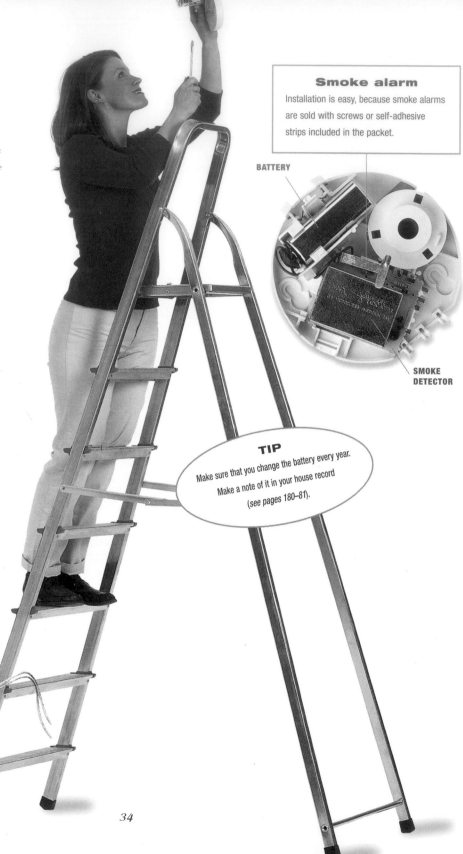

Smoke alarm

Installation is easy, because smoke alarms are sold with screws or self-adhesive strips included in the packet.

BATTERY

SMOKE DETECTOR

TIP

Make sure that you change the battery every year. Make a note of it in your house record (*see pages 180–81*).

FITTING A NEW MORTISE LOCK

Most external doors, especially a front door, will have at least one lock, usually a cylinder rim lock (the type that will lock automatically when the door closes), but it is a good idea to install an extra mortise lock for added security. Also, if you have moved into a new flat or house, you may want to change the locks simply for your own peace of mind. In this case you can just remove the old lock, then easily customize the existing recess to fit the new lock.

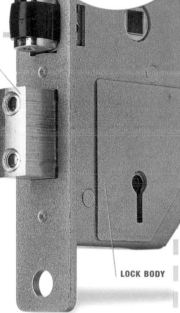

FACE PLATE

BOLT

LOCK BODY

FITTING THE LOCK

TAKE YOUR TIME!

In general it is advisable to fit the lock approximately one third of the way up the door's length in a part of the door that is solid.

need

* new lock
* pencil and bradawl
* drill plus bit
* chisel
* hammer
* pad saw
* screwdriver

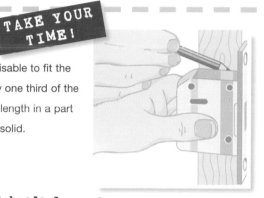

1 Place the lock on the door edge in the desired position and make pencil lines to mark the exact size of the mortise. Mark the exact centre of the door vertically between the first two pencil lines.

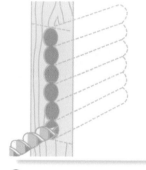

2 Using a drill fitted with a bit that is the diameter of the width of the lock, drill a series of holes very close together along the centre line you have marked. This will create the recess for the mortise.

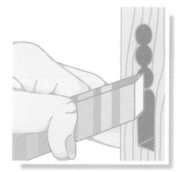

3 Chisel the holes into a rough rectangle, then chisel a little more carefully to make a clean recess for the lock body. Next, chisel a wider shallow recess only a few millimetres deep for the lock plate.

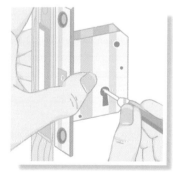

4 Use the new lock as a guide to mark the position of the keyhole with a bradawl. Drill the hole and cut it to a more exact shape with a pad saw. Insert the new lock into the recess, then screw in place.

5 With the lock bolt protruding, close the door and mark the position of the striking plate on the door frame. Drill and chisel a recess for the bolt box, then a shallow recess for the striking plate. Screw in place.

6 Double-check now that the new lock works properly and that the door opens and closes easily. To finish, screw a decorative key plate over the keyhole on both sides of the door.

Having problems with your waterworks?

PLUMBING ETC.

Any part of your plumbing that has moving parts will sustain wear over time. Limescale build-up is a real nuisance; rubber washers or seals will need replacement at some time and toilets need a certain amount of maintenance over their lifetime to ensure they work every time! Spare parts for plumbing repairs are usually inexpensive, so by doing it yourself you can save money.

DRIP, DRIP, DRIP

A constantly dripping or leaky tap can be a great source of annoyance. If water leaks from the spout, then it is usually the result of a worn, old or faulty washer. You can easily fix this with no more than a screwdriver and a spanner (and a new washer, of course). Take a look at your taps, for there are a variety of styles ranging from the traditional pillar tap to the more modern shrouded-head type.

✻ I LOVE MY RUBBER DUCK, BUT A WASHER IS MORE USEFUL!

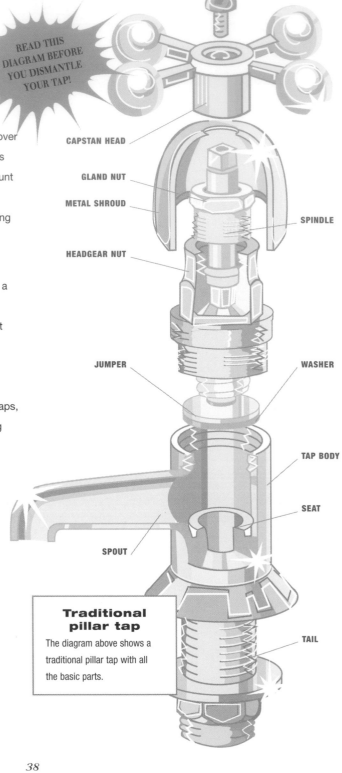

READ THIS DIAGRAM BEFORE YOU DISMANTLE YOUR TAP!

CAPSTAN HEAD

GLAND NUT

METAL SHROUD

SPINDLE

HEADGEAR NUT

JUMPER

WASHER

TAP BODY

SEAT

SPOUT

Traditional pillar tap

The diagram above shows a traditional pillar tap with all the basic parts.

TAIL

CHANGING A TAP WASHER

12

A screwdriver, spanner or wrench and a new washer is all that's needed for this task.

CHANGING A WASHER ON A MODERN SHROUDED-HEAD TAP

Modern taps usually have the head and cover in one piece, which acts as the on/off handle. This must be removed first to expose the headgear nut. Most covers are fixed with a single retaining screw that is quite easy to undo, while others don't have a screw at all and can be pulled from the headgear, which is even simpler. If the water seeps from the head of the tap (the bit that you turn), then the 'O' ring may need replacing. Turn off your water supply and carefully disassemble the tap. Remove the rubber 'O' ring and replace with a new one.

need
* adjustable spanner
* screwdriver
* new washer

1 First turn off the water supply. Remove the tap handle/cover by gently prising off the hot/cold indicator to reveal the screw inside.

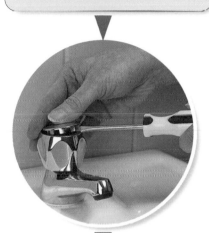

2 Loosen the screw and remove the cover to reveal the headgear. The shrouded head cover which forms the handle will slip off easily, exposing the top of the brass headgear inside.

3 Use an adjustable spanner or a suitably sized spanner to remove the headgear from the body of the tap. The washer is located at the base of the headgear. It may be attached to the jumper or it may be lying inside the body of the tap itself. If attached to the jumper it will be pressed over a small button in the centre.

4 Simply prise the old washer off with a knife or screwdriver and replace it with a new one of the same size. (Take the old one to the plumbing supplies shop and buy one that is the same.)

AND A PILLAR TAP

Remove the cover to reveal the headgear nut. Unscrew and remove the headgear using a spanner, then remove the assembly. The washer is located at the base of the assembly, attached to the jumper. Prise off the old washer using a knife or a screwdriver, and then fit a new one.

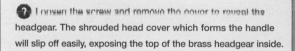

'O' RING   HEADGEAR   WASHER

Toilet troubles and ballcock blues

CHANGING THE TOILET SEAT 13

A toilet seat is a very personal thing. When you move into a new flat or house, it's probably one of the first things you'll want to fix or change. This is a very simple procedure that will make you very happy indeed. There is a huge selection of weird and wacky toilet seats available now, so choose something tastefully tame or wonderfully wild to suit your bathroom or your mood!

1 First locate the seat's screw attachments underneath the toilet at the back of the bowl. Unscrew and then remove the old seat and bolts.

2 Your lovely new toilet seat will undoubtedly arrive complete with installation instructions. Simply insert the bolts as directed and snap or screw on the new seat. At this point it is important to 'test drive' your new seat, just to make sure that it doesn't wobble and is in the correct position.

3 Now stand back and admire your wonderful new 'throne'!

Toilet seats this glamorous will never be left in the 'up' position!

❋ YOUR TOILET SEAT SAYS AS MUCH ABOUT YOU AS YOUR PERFUME DOES!

MENDING OR ADJUSTING THE BALLCOCK

14

Most toilets have a direct-action cistern. The action of the ballcock (or float) opens and closes a valve, allowing water into the cistern. When the toilet is flushed, the water empties, the ballcock drops and the valve opens. The cistern fills, lifting the ballcock as the water level rises. When the correct level is reached, the valve is closed and the water supply stopped. Sometimes the float arm needs to be adjusted to achieve the optimum level of water in the cistern. Bend the float arm downwards slightly to reduce the water intake or straighten it out a little to increase the intake. A floatless ballcock has an adjusting screw with a locknut. Release the locknut and screw the adjustor towards the valve to decrease the water level and away from the valve to increase the level.

HEY **LEARN YOUR WAY AROUND YOUR CISTERN**

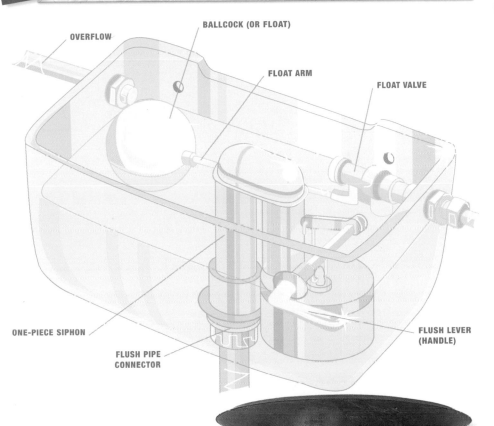

- OVERFLOW
- BALLCOCK (OR FLOAT)
- FLOAT ARM
- FLOAT VALVE
- ONE-PIECE SIPHON
- FLUSH PIPE CONNECTOR
- FLUSH LEVER (HANDLE)

UNBLOCKING THE TOILET

15

A blocked toilet is a very unpleasant thing. However, here are a few things that you can easily do to rid yourself of the problem. If the water rises or drains away slowly when you flush the toilet, this means that there is a blockage or partial blockage in the trap. You now need to remove the blockage. Obtain a large plunger from a hire shop – this looks like a sink plunger but is much larger. Place the narrow part of the plunger down inside the bend at the base of the toilet, grip the handle firmly and pump vigorously a few times. Eventually the blockage will clear and the water level will drop to normal.

If this fails, it could mean that the trap is more solidly blocked. You now need to hire a WC auger, which is a specially designed flexible rod with a crank handle at one end. Place the flexible part into the trap, then crank the handle. This should dislodge the blockage and cure the problem.

REMEMBER
Disinfect rods and plungers thoroughly before returning them to the hire shop.

Faulty flushes and leaky baths

MENDING THE TOILET FLUSH

A toilet that doesn't flush regularly or requires a 'knack' may be the first and most annoying thing you will experience when you move to a new flat or house. The previous inhabitants probably got used to it that way, but you don't have to – just follow the next three steps to fix it!

EASY!

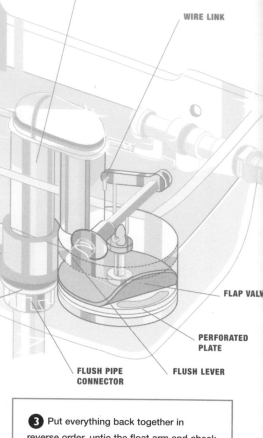

ONE-PIECE SIPHON

WIRE LINK

SEALING WASHER

RETAINING NUT

FLAP VALVE

PERFORATED PLATE

FLUSH PIPE CONNECTOR

FLUSH LEVER

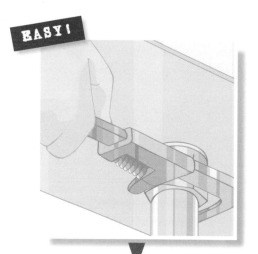

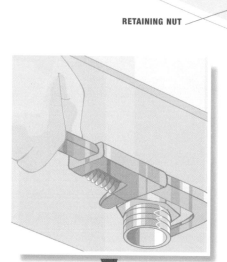

❸ Put everything back together in reverse order, untie the float arm and check the flush. If your flush arm is loose or won't work at all, simply check the wire link at the end of the flushing arm. Tighten if loose or replace if broken – you can make a new one from thick wire.

❶ First check the water level and flushing mechanism. If these are fine, then you will probably need to change the flap valve that releases water into the toilet bowl. Flush the toilet, then tie up the float arm so that the cistern does not fill up. Underneath the cistern is a large nut that holds the flush pipe. Use a wrench to release the nut.

❷ Remove the flush pipe connector and then push it carefully to one side. Loosen the retaining nut that holds the siphon to the inside of the cistern. Disconnect the flush arm from the siphon and carefully ease the whole assembly out of the cistern. Lift the rubber flap valve off the perforated plate and then replace it with a new one of the same size.

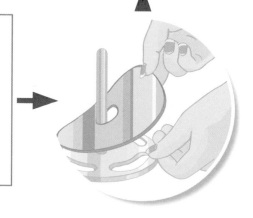

* NOTHING LIKE A POWER SHOWER TO SET ME UP FOR AN EVENING'S DIY!

FIXING A LEAKY PLUG

Sink and bath plugs also do not last for ever. Over time the rubber shrinks and as a result lets water out when you need to keep it in. There's nothing quite as relaxing as a long soak in a hot bubble bath (especially after a strenuous day of DIY), but not when the water seeps out and you have to constantly add more to the bath.

All the bits and pieces required to complete the task are inexpensive and readily available at DIY stores. You can get a link plug chain if preferred, but I rather like the one shown here.

END CLIP   A-RING

FIXING A SHOWER HEAD

Shower heads don't last for ever, and unless you're lucky enough to be able to afford to install a new suite, you are likely to have to make do with a little 'quick-fix' procedure while you save up. Limescale build-up may cause a problem inside the shower head when the perforations in the rose become clogged and restrict the water flow. Also corrosion of the head or metal hose may cause water to leak and spray everywhere. Unscrew the head and/or the hose and discard it. Buy new ones from the DIY store and simply replace them. Not quite the same as a new power shower, but it may just be a bearable stopgap.

1 Get a new plug, the same size as the old one, plus any extra items such as A-rings and end clips (*see photograph*). Next, use a pair of long-nosed pliers to attach the new plug to the bath or sink. If you forget to take the old plug to the DIY store, just purchase a universal 'one-size-fits-all' one that will do the job, but they are easier to displace.

2 Slip the end clip onto the last ball of the chain and then pass the A-ring through the holes in the clip. Then simply attach it to the plug.

3 If the plug chain has snapped, which happens frequently, just take off the old chain and add a new length using the same fixings.

* NOW THE PLUG'S BEEN FIXED, NO MORE HANGING OUT IN THE BATHROOM WITH THE LADS.

All about the washing machine

PLUMBING IN A WASHING MACHINE/ DISHWASHER

19

I couldn't possibly live without my washing machine – a sentiment I think we all share. All machines have comprehensive instructions for installation, but basically it requires a hot and cold water supply. At the back of the machine you will see two coloured rubber hoses, which are red for hot and blue for cold – simple. You need to connect these hoses to your household plumbing system; self-boring valves have been designed especially for this purpose. However, connection works only if the water pipes are in a convenient position, i.e., running behind or alongside the position of the machine. Re-routing of pipework is a professional job.

✱ DON'T RUN, THE JOB'S DONE!

HEY BEHIND THE WASHING MACHINE

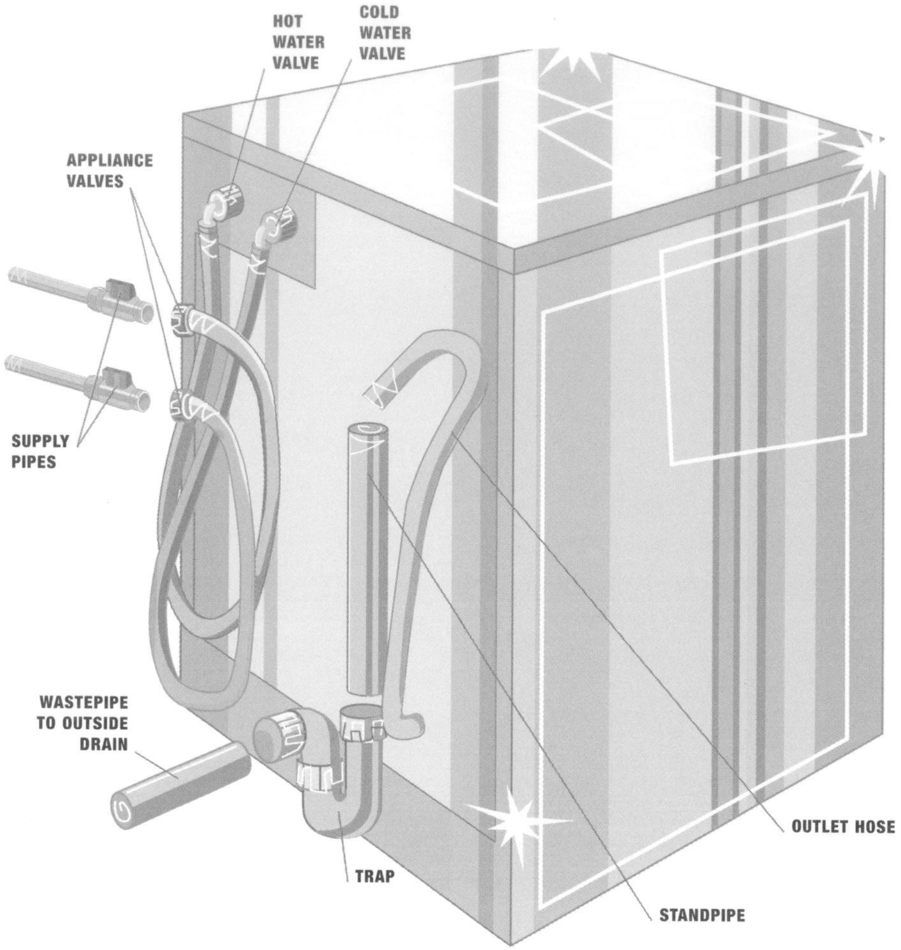

HOT WATER VALVE

COLD WATER VALVE

APPLIANCE VALVES

SUPPLY PIPES

WASTEPIPE TO OUTSIDE DRAIN

TRAP

STANDPIPE

OUTLET HOSE

DON'T PANIC!

Plumbing in a washing machine shouldn't be like a task from *Mission Impossible*! Just take a look at the diagram above. It identifies all the bits of spaghetti dangling from the back of the machine, showing you the essential 'ins' and 'outs' of your trusty washing machine. If possible, try to situate the machine next to an existing sink – it really will make life easier, because you'll have all the waterworks you need and proper functional drainage facilities close by. Hot and cold water supply pipes will be colour-coded so there'll be no mistakes made – or will there?

CLEANING THE WASHING-MACHINE FILTER

20

When you purchase a household appliance, it is a good idea to keep any proof of purchase such as receipts, along with the installation instructions and operating/maintenance manuals, in a safe place. These will advise you of any maintenance procedures to follow and what to do should a simple problem occur. Some machines have a filter that removes all sorts of flotsam and jetsam from the washing water. It will, of course, eventually clog up – hair bands, small change, safety pins, you name it. So check the filter regularly and remove any foreign bodies that are lurking inside. The manual will show you where it is and how to open it. Some machines don't have an integral filter but have a non-return valve in the waste-water outlet. The function of this is self-explanatory: it lets water out but not back in again. This can also become clogged with lint and other bits and pieces. Result: the water is not expelled and stays in the machine. Action: turn off the water, pull out the machine, locate the valve, unscrew, unclog, then put everything back together.

DON'T PANIC!

If your machine appears not to be washing properly, you should ensure that the filters are clean. To do this, first turn off both the water and the electricity supplies. Next unscrew the hot and cold water inlet hoses from the back of the machine, and you'll see the filters. They are made of a fine mesh, which from time to time may become clogged. Pull out the filters with a pair of flat-nose pliers, and clean them under a running tap. Replace the filters and screw the inlet hoses back onto the machine. You only need to tighten the hose connections firmly by hand to avoid any leaks.

❋ I JUST KNOW HE'S BOUGHT ME A LOVELY NEW FILTER!

WARNING

❋ Before undertaking any maintenance on your washing machine, however minor, always ensure that you turn off the electricity supply to the machine.

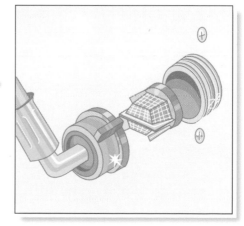

Quick and easy fixes for the kitchen

CHANGING THE DIRECTION OF THE REFRIGERATOR DOOR

Every modern refrigerator offers this option. It's very simple and can make life in your kitchen a lot easier. Check the appliance manual for instructions – each refrigerator will be a little different, but the basic principle is the same. Simply undo all the fixings on one side and move them to the opposite side, where all the holes are pre-drilled for you!

1 Unplug the refrigerator and pull it away from the wall slightly (or, if it's a new refrigerator, change the door direction before installation). Carefully lean the refrigerator back so that you can see the metal hinge plate on the bottom of the door. Unscrew the plate at the base, then remove the door.

2 Now unscrew the fixing at the top of the hinge side and install it in the same position on the other side – there will be holes there already for this purpose. Remove the plastic hole covers. Reattach the refrigerator door and replace the metal plates at the base.

3 When you are satisfied that the refrigerator door is secure, locate those little plastic caps that you just removed and insert them in the empty screw holes on the original hinge side of the door, so that it looks neat.

CHANGING THE BULB IN A REFRIGERATOR

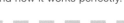

The previous owners of my flat were kind enough to leave me their refrigerator and freezer, and as I was completely penniless, it was most gratefully received. However, I couldn't quite work out what was wrong with it. On the day after I moved in I realized that there was no light inside – no wonder I had difficulty finding anything.

To change the bulb, simply locate the bulb housing, remove the bulb, take it to the DIY store, buy one that's the same and replace it. Easy.

I experienced a similar problem with my ancient gas cooker – the rings didn't light automatically. On closer inspection I found that there was a battery in the bottom compartment. All I had to do was replace the battery and now it works perfectly.

FIXING DRAWER RUNNERS

23

Yet another 'make do and mend' exercise. Drawers that stick or drop out when you open them are enough to make you want to tear your hair out. If this is happening in your kitchen – keep calm! Simply remove the offending drawer and take a look at the drawer runners to see what the problem is. More often than not the screws that hold the drawer runner to the cabinet carcass will have worked themselves loose due to many openings and closings. This is very common, especially if the cupboards are the self-assembly types that have seen more than their fair share of wear over the years. The runners themselves are usually made of plastic or metal. Make sure that all the screws are tight, and replace any that are missing. If the screw holes are damaged or have become a bit too big for the screws, just use larger screws, which should hold the runner securely in place. Replace the drawer, then test to make sure that everything is running smoothly.

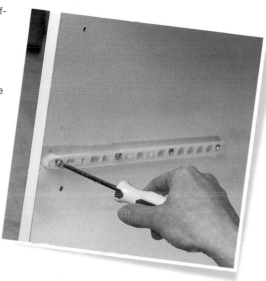

ADJUSTING HINGES ON KITCHEN CUPBOARD DOORS

24

Over time, with constant opening and closing, the hinges on kitchen units wear, become loose and can fall out of alignment with the carcass, causing many difficulties. You need a couple of screwdrivers to solve the problem.

1 Open the doors and take a look at the hinges. You will see that the door part of the hinge is attached with a large screw to the mounting plate on the carcass. By tightening or loosening this screw you will be able to move the door inwards or outwards.

2 If your doors are not lining up properly when closed, for side-to-side adjustment, loosen or tighten the smaller screw as shown. A little patience will ensure that the doors line up with one another and that they open and close properly.

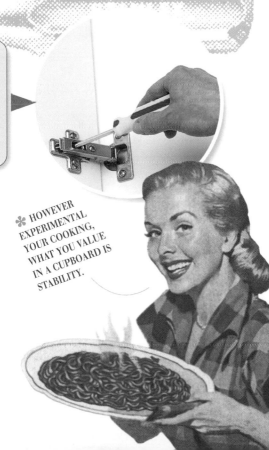

❋ HOWEVER EXPERIMENTAL YOUR COOKING, WHAT YOU VALUE IN A CUPBOARD IS STABILITY.

Hold your nose — it's time for the stinky stuff

BLOCKAGES

Well, it had to happen eventually. It's time to slip on those fabulous rubber gloves and get down to clear the drains. Yes, it may be unpleasant, but sometimes you just have to bite the bullet and do it. Blockages and the like occur from time to time for a number of reasons – it's not life-threatening, just a bit unpleasant, that's all.

UNBLOCKING GUTTERING

25

Gutters, downpipes and hoppers can easily become clogged up with leaves or other debris. In time this will result in an overflow, which can cause damage to brickwork. You'll have to put on the rubber gloves and climb up a ladder. Clear the debris from the guttering and hopper by hand, then probe the downpipe with a long stick or cane to force out the blockage.

A girl's gotta do what a girl's gotta do, but she can do it with style.

UNBLOCKING DRAINS

26

Blocked drains are a nuisance, but if you can remember to check them regularly, they shouldn't become a problem. During autumn, leaves and other organic matter tend to clog up drains and cause overflows – time to put on those rubber gloves again!

1 If the blockage occurs in a yard gully, put on protective rubber gloves, remove the metal grid and put your hand in. If the drain is overflowing with water, bail some out first, then scoop out the debris from the trap until the water drains away.

2 When the blockage is cleared, rinse the gully with fresh water and disinfectant. Afterwards scrub clean the metal grid before replacing it.

3 If the soil stack that takes toilet waste to the main drain is blocked, you should consider calling in the professionals, as it is not a pleasant or easy task. Additionally, if you're not good with heights, then it may be better to bring someone in.

TIP
It is a good idea to make a note of this kind of maintenance in your house record (*see pages 180–81*). Prevention is better than cure.

CLEARING THE S-BEND

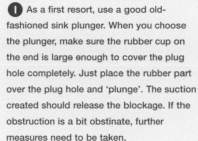

Blockages in sinks can occur quite frequently, especially if you are a bit careless about what you try to wash down the plug hole. It is not a good idea, for example, to pour hot candle wax down the sink (I learned this the hard way while doing a spot of candle-making) because the wax solidifies and causes a fine blockage indeed. This can also happen with cooking oil, so beware. If you do have a blockage, the following steps will show you how to get things moving again.

1 As a first resort, use a good old-fashioned sink plunger. When you choose the plunger, make sure the rubber cup on the end is large enough to cover the plug hole completely. Just place the rubber part over the plug hole and 'plunge'. The suction created should release the blockage. If the obstruction is a bit obstinate, further measures need to be taken.

2 Take a look under the sink; you will see something like this. The wastepipe will be attached to a trap of some sort. It may look more like a bottle, in which case you can just unscrew the base to gain access to the pipe. Place a small bowl under the trap, then unscrew and remove the trap. Pour out any water that may be caught inside.

3 Use a piece of wire to clear out the blockage – a straightened-out coat hanger is perfect for this purpose. Bend a little hook on one end to facilitate removal of the obstruction. Clean the trap with water before replacing it. Make sure you use another sink, though, or you'll find water all over the floor!

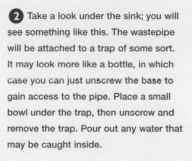

need

* sink plunger
* small bowl
* thick wire or straightened-out coat hanger

I DO THE ROUGH STUFF ... HIS NOSE IS SO MUCH MORE SENSITIVE THAN MINE.

4 If you still have the nasty blockage, try probing the branch pipe with the wire and almost certainly the problem will be resolved. Reassemble the trap, then flush with fresh water and disinfectant.

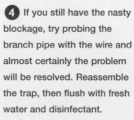

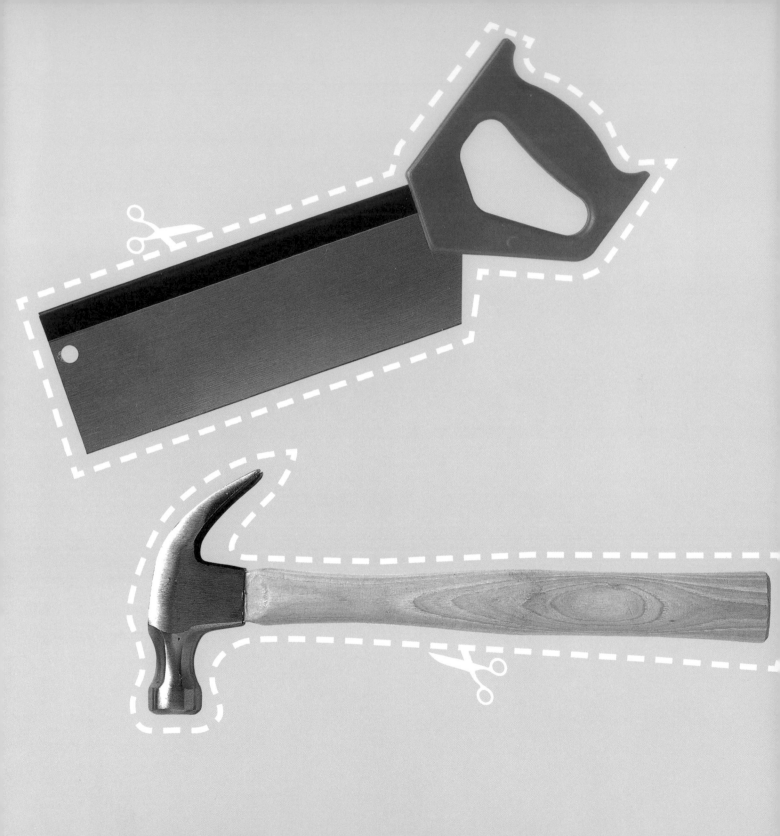

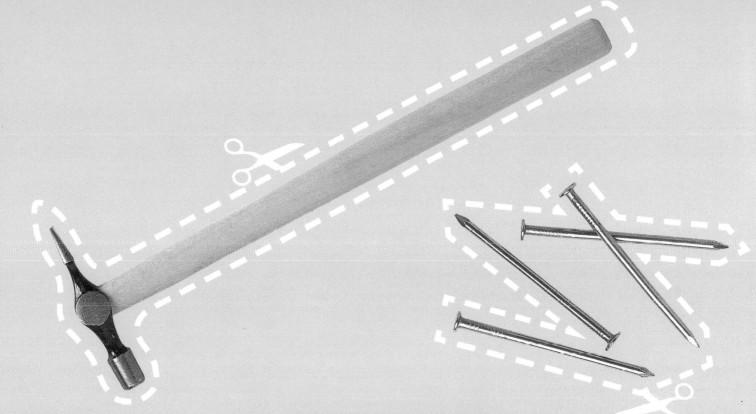

LAYING
IT

It's time to find out what you've been

walking on all these years!

Flooring and how to lay it

FLOORING CAN be divided into two very basic categories: 'hard' and 'soft'. Which avenue you choose to pursue is entirely a matter of personal preference. In the hard-flooring category I'm going to include floorboards, woodstrip flooring, floor tiles (ceramic, quarry, terracotta and marble) and parquet flooring. The soft-flooring category includes carpet, carpet tiles, sheet vinyl, vinyl tiles and cork tiles. Hard flooring should be able to withstand more punishment, whereas soft flooring is more appropriate in the bedroom.

CHOOSING IT

To a certain extent the type of flooring you choose will depend on the health of your bank balance (or credit card!), and of course the location of the flooring is a very important concern. For example, bathrooms, kitchens and utility rooms suggest a vinyl, tile or wood-flooring approach, whereas the bedroom or living room requires the soft carpet or floorboard and fluffy-rug touch. Traffic is another consideration – do you have hoards of people trundling up and down your hallway? If so, a hard-wearing carpet or easy-to-clean tile or wood floor is an option. What your heart desires sometimes has to be tempered by what you already have – beautifully polished old wooden floorboards are lovely, but if you have a concrete subfloor under the scruffy carpeting, then perhaps it's time to compromise. It really is worth spending a little time considering what you want, what you've got, what you can live with and what you absolutely have to get rid of right away. Do you have the time and funds available for a fairly big project or is a quick-fix option more attractive? A question of assessing the tolerance levels, I suppose. Making the wrong choice in haste can be costly and ultimately disappointing – there is so much to choose from out there, so do take your time and get it right.

✳ CREATE YOUR OWN FABULOUS DANCE FLOOR!

LAYING A BRAND-NEW SUSPENDED FLOOR IS A BIG PROFESSIONAL JOB (AND EXPENSIVE), BUT WOODSTRIP, A NEW CARPET OR POSSIBLY A GOOD-QUALITY VINYL PLUS RUG ARE FAIRLY ACCESSIBLE OPTIONS.

REPAIRING A FLOORBOARD

So having dispensed with the unsightly, worn-out carpet, you now have to decide what to do next. Having taken away the carpet, assess the state of the floorboards. If you're very fortunate, then all the boards will be perfect and you can happily skip to the next section. If, however, some of the floorboards are either loose or damaged, then you'll need to repair or replace them. Until the original flooring has been removed, what lies beneath is always a mystery. There may be rotten boards, wobbly ones, multicoloured ones, squeaky ones – you just never know.

1 Take your bolster chisel and hammer and begin at a join over a joist (the butted ends of floorboards usually meet above a joist). Hammer the broad end of the chisel into the join and start to lever up the board. This may take a little bit of effort at first, but persevere – it will come loose.

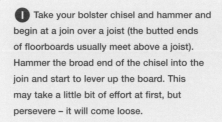

2 When the end of the board has been loosened, take a piece of thick cardboard and position it over the end of the next board to protect it from damage. Prise the floorboard up using a claw hammer. Remove the board and any nails that have pulled through the board and remain in the joist.

3 Brush away any debris that may be lying on the joist. Cut the new floorboard to size and lay it in position. The new floorboard is unlikely to be the same width as the old one, so fill in the gap with a thin wooden lath cut to the same size. Fill in any other large gaps in this way too. Hammer in flooring brads to secure the boards to each joist.

TIP
When you purchase the new board, make sure that it is the same thickness as the old one so the surface of the floor remains even.

4 Having replaced all the damaged boards, it's now time to fix the loose or squeaky ones. A loose or squeaky board has probably worn or warped over time and has worked itself loose from the joist. This creates the wobble or a squeak when walked on. Lift the board in the same way, then remove old nails, debris and so on. Replace the old board in the same position, then use screws at least twice as long as the board is thick to secure it to the joists. Use the old nail holes or drill new holes if the old ones are too worn.

need
* bolster chisel
* claw hammer
* cardboard
* new floorboard
* lath
* saw
* hammer
* flooring brads

No floor is complete without a bit of skirting!

SKIRTING BOARDS

29

Skirting boards serve both a functional and a decorative purpose. On the functional side, a skirting board protects walls from accidental damage or scuffing from passing pedestrians, while decoratively they can add the perfect finishing touch to your decor. It may be necessary to replace damaged skirting boards or, more often than not, the style you have inherited with your new home looks too dated for the decorating plans you have. Installing new skirting boards is relatively easy. They can be attached to plasterboard with finishing nails or to masonry walls with masonry nails. The only difficulty you may face is making them fit around external and internal corners: a little tricky maybe, but not impossible.

need

* small crowbar
* skirting boards
* pencil
* jigsaw or coping saw
* wood glue
* hammer
* masonry nails or lost-head nails
* wood filler
* sandpaper

EXTERNAL CORNERS

1 To begin, use a small crowbar to prise the old skirting boards away from the wall. Remove any nails that may be protruding from the wall. For an external corner you will need to make mitre joins for a neat finish. Measure the first board accurately and cut the end to 45 degrees using a jigsaw with an adjustable sole plate.

2 Hold a length of skirting board against the other wall. At the corner, mark the profile of the mitred edge of the first board on the back of the second one using a pencil. Return the board to the workbench and carefully cut another 45-degree angle along the pencil mark, using the jigsaw as before. Offer up the board to the corner again to check the fit and make any necessary adjustments before fixing it into place permanently.

3 Apply some wood glue to the cut mitres when you nail the board to the wall. Wipe away excess glue from the surface of the wood using a damp cloth. If there are any small gaps, it is simple to fill them in with wood filler, then sand smooth when dry. You can fill in any screw or nail holes in this way too, then retouch any paint.

* REMEMBER, SON, SKIRTING IS BEST LEFT TO THE GIRLS!

INTERNAL CORNERS

RIPPING OFF OLD SKIRTING BOARDS IS A GREAT STRESS RELIEVER!

1 Begin by cutting the first length to fit along a straight wall. Butt the ends squarely against the walls. If you need to join lengths together, do so using wood glue. Fill any gaps with wood filler, then sand smooth and retouch with paint later. Don't fix the first board to the wall yet because you will need to remove it later in order to cut the end to shape.

2 Using an offcut of the skirting, trace the profile on the end of the next piece to be fitted. Measure the length of the board and mark it with a pencil line – you will need to cut it straight like the first one. However, don't do it yet in case you make a mistake cutting the profile and have to start again.

❋ NOW THAT'S A NICE BIT OF SKIRTING!

TIP
I thought it was quite a good idea to paint the skirting boards before fitting – that way you don't have all that painting to do afterwards.

3 Cut carefully around the traced profile, using a coping saw or a jigsaw. It may take a few attempts to make it fit perfectly, but keep trying. Check the fit of the board and recheck that the length is correct; then cut the end straight across.

4 When you are confident that the second piece fits the first correctly, place it in position and nail it to the wall securely. When you return to the first board, remove it and scribe the shaped profile using an offcut as before; then cut it out carefully so the end fits neatly over the last board.

Stripping with all your clothes on

SANDING A FLOOR

Sanding a wooden floor is a big job, but doing it yourself is a much cheaper option than hiring someone to do it for you. Large sanders and edge sanders are readily available from hire shops, most of which have good rates for weekends or even cheaper midweek deals. The equipment comes with complete instructions, and the shop will sell you plenty of sanding disks for the edger and sheets for the large machine; usually they'll give your money back for the ones you don't use. Most hire-shop staff are very helpful and will talk you through the basics.

1 Preparation pays! Check the floor and fill any gaps, secure loose boards and repair any damaged ones. Now take the nail punch and hammer and sink all the nails well below the surface of each board. Yes, a little bit laborious, but it really is important that no nail heads protrude because they will tear the sandpaper. Finally, brush any debris or loose paint and so on from under the skirting boards.

2 Having completed all the preparations, the diagrams below show you how to order your work. Take note of the directional arrows.

3 When all the sanding has been done, vacuum the floor thoroughly, then wipe it down with a cloth soaked in white spirit. See the next section for a few finishes that you might like to apply to your newly sanded floor.

ORDER OF WORK

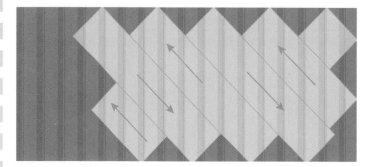

☞ Use the large sanding machine with coarse abrasive paper in a diagonal direction across the floor. This first procedure evens out the floorboards and removes the top layer of dirt. Work across the floor again, but this time diagonally the other way.

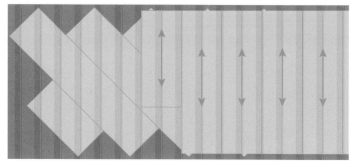

☞ Next, work in strips across the boards. With the machine on, sand in one direction, and then work back along the strip. Switch off the machine and move to the next strip; repeat. Work first with coarse sandpaper, then medium sandpaper, then fine to finish.

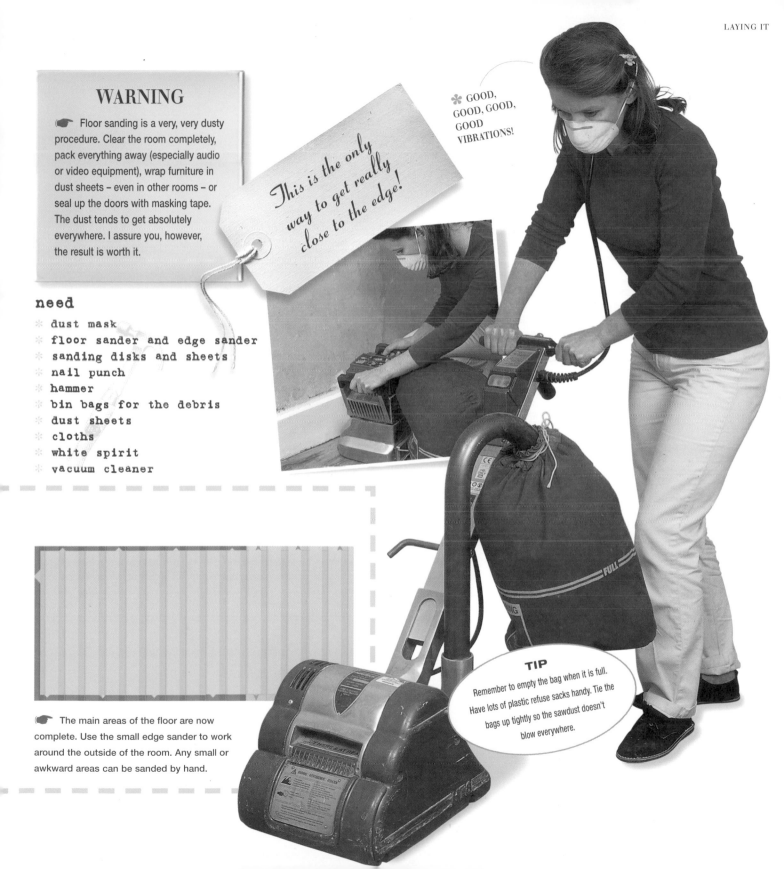

WARNING

☞ Floor sanding is a very, very dusty procedure. Clear the room completely, pack everything away (especially audio or video equipment), wrap furniture in dust sheets – even in other rooms – or seal up the doors with masking tape. The dust tends to get absolutely everywhere. I assure you, however, the result is worth it.

This is the only way to get really close to the edge!

✱ GOOD, GOOD, GOOD, GOOD VIBRATIONS!

need
* dust mask
* floor sander and edge sander
* sanding disks and sheets
* nail punch
* hammer
* bin bags for the debris
* dust sheets
* cloths
* white spirit
* vacuum cleaner

☞ The main areas of the floor are now complete. Use the small edge sander to work around the outside of the room. Any small or awkward areas can be sanded by hand.

TIP
Remember to empty the bag when it is full. Have lots of plastic refuse sacks handy. Tie the bags up tightly so the sawdust doesn't blow everywhere.

Adding that essential 'touch of glamour'

FINISHING OFF

Freshly sanded wooden floorboards look fantastic, but there's still a little more work to do before you can put your feet up! Bare wood must be protected in some way, either with a suitable sealant or varnish. Alternatively, you might like to use paint or a coloured stain. If you feel a little creative rush, take a look at the decorative flooring ideas on pages 76–77. Remember, any decorative paint or stain effect must be sealed with a few coats of clear varnish on completion or it will soon be worn away. Finally, a little advice about the order of work: begin in the corner furthest from the door and work your way back. Do not paint yourself into a corner!

VARNISHING FLOORBOARDS

31

Applying varnish to floorboards is in essence the same as varnishing any other wooden surface. The advantages of varnishing are twofold: first, to protect the wood from damage and staining and, second, to enhance the natural beauty of the wood grain. Choose a clear natural varnish or a tinted one to change the colour of the wood. There are, however, a few safety considerations to take into account. Some varnishes or sealants are water-based and low-odour, but the solvent-based ones can give off fumes and are unpleasant to use if working on a large area. Remember to open all windows to keep the room well ventilated.

> ALWAYS WEAR GLOVES IF YOU WANT TO PROTECT YOUR HANDS FROM UNSIGHTLY PAINT STAINS.

① Use a good-quality, spotlessly clean paintbrush to apply varnish, preferably one that has not been used for any other kind of painting, since old paint particles are likely to spoil the finished appearance of the floor. Varnish may also be applied with a small roller in order to save time.

② When using a brush, dip about one third of the length of the bristles into the varnish, then gently touch the end of the brush against the rim of the tin to remove the excess. Try not to stir or wave the brush around too much; this may create air bubbles that will spoil the finish just as dirt or dust particles would. If this happens, however, don't despair; simply rub over the area with fine sandpaper. Remember also to brush out the varnish well as the work progresses to prevent it from collecting in pools.

✳ AND THERE'LL STILL BE TIME TO TOUCH UP YOUR NAIL VARNISH AFTERWARDS!

PAINTING FLOORBOARDS 32

Floor paint is one way of decorating a wooden floor if a solid colour is preferred. Many paint manufacturers produce paint that can be used on wooden, concrete and brick floors, producing an extremely hard-wearing glossy sheen. This is ideal for bathrooms and kitchens where wipe-clean properties are a big advantage, or in areas where there is likely to be more-than-average wear and tear. Choose from a wonderful range of fresh colours to complement your furnishings.

BLEACHING FLOORBOARDS 33

Wood bleach will lighten stained wood and in most cases bring it back to the original colour. Wooden floors that have darkened with age can be given a facelift by bleaching, but remember to use a varnish stripper first to remove the old finish. Alternatively, wood bleach can be used to remove stains caused by accidental spills, or perhaps if the original floorboards are unevenly coloured. Following the manufacturer's instructions, apply the wood bleach with a brush, either to the area of the spill or stain or, preferably, to the whole floor to avoid patchiness. The floor can then be sanded and the chosen finish applied. Mild staining can be bleached out by applying a very strong solution of hot water and oxalic acid crystals. The crystals can be obtained from a good pharmacist. Rinse the floor surface with clean water and always use protective gloves.

STICKY

1 Apply the floor paint with a small long-handled roller. This saves time and also your back if the floor area is large. Use a small brush to paint around pipes and to reach any small awkward places.

2 Remember not to overload the roller – if paint seeps into and dries in the spaces between the boards, it can look unsightly. It is far better to apply three or four thin coats, allowing each to dry before applying the next. If you use a water-based, quick-drying floor paint, it is quite possible to complete painting your floor in a day.

STAINING FLOORBOARDS 34

Floor stain colours the wood but does not disguise the grain in any way. It is not always possible to visualize the effect of the chosen stain just by looking at the picture on the tin. Experiment on a small unobtrusive part of the floor first, before you take the plunge with the whole area. There are some beautiful delicate stains on the market at the moment, as well as stronger, more intense shades. Just take your pick.

1 Apply wood stain or wash using a sponge or soft cloth. Several coats can be applied to intensify the colour or shade. Always allow the first coat to dry before applying the next.

2 Work with the grain of the floorboards in parallel strips, trying not to overlap the strips as you go. When you're satisfied with the shade, apply two coats of clear matt or soft sheen varnish to seal it.

So many floor coverings, so little time

CHOOSING FLOOR COVERINGS

35

A girl could indeed have her mind completely boggled by the array of different types and styles of flooring products available. The recent trend for polished wooden floors and woodstrip flooring has not rendered carpets obsolete in any way, though there is an increasing availability of natural-fibre woven floor coverings such as jute or seagrass. The first basic consideration to be made is how much you've got to spend; the second is the look you intend to achieve; and the third is the amount of wear the flooring will be subjected to. The traditional choice is carpet for comfort zones like bedrooms and lounging areas, and more practical surfaces for other areas. But ultimately the choice is yours. Here are a few examples.

Carpet tiles

These are generally thought of as flooring for offices, hotels or commercial properties because of their resilience. Do not be put off by their rather drab track record – carpet tile companies produce a fine range of colours these days and this type of flooring can be a smart choice for the home. Carpet tiles can be taken up and replaced if stained or damaged, or even rotated and swapped around to spread out areas of wear. They are easy to cut and shape with a craft knife. There is also the option of creating chequerboard effects, contrast borders or even a carpet mosaic if you feel adventurous.

Estimating the quantities

For floor tiles, calculate the size of the floor by multiplying the length and width measurements of the floor. Each packet will give the approximate area that the contents will cover. Divide the floor area by this figure to calculate how many packets you'll need. For sheet vinyl or carpet, measure the room and take a scale plan to the flooring supplier, who will advise you. Carpet and vinyl come in various standard widths; if the room is large or irregularly shaped, the material may need to be joined to avoid waste.

Woodstrip flooring

This type of flooring is commonly referred to as 'floating' flooring. Solid wood or the less expensive laminated planks are made in a tongue-and-groove style, designed to be glued and then slotted together. The resulting floor sits on top of the subfloor and in effect 'floats'. Woodstrip flooring is immensely popular these days for obvious reasons: it's not hugely expensive, it looks fabulous and it's a breeze to keep clean. Get some!

Carpet

Always a popular choice, good-quality carpet made from natural fibres can last a lifetime if well cared for, but it can also be expensive. Cheaper versions made from synthetic fibres or a blend are easier on the pocket, but less durable. They are perfectly acceptable, however, for a small room that will not receive rough treatment.

Parquet flooring

Looks like large square tiles. It is in fact a thin layer of decorative wooden strips or little squares laid in a block or in herringbone-type patterns, then glued together to form easy-to-handle tiles. Some parquet flooring tiles are self-adhesive, but others must be glued to the subfloor.

Cork tiles

Similar to vinyl tiles in that most are self-adhesive or can be attached with contact adhesive and are easy to cut and fit. They are usually made from cork granules that are compressed and stuck together under heat and pressure. Some can be purchased ready finished, or you can varnish them yourself. Cork flooring is quite resilient and is warm underfoot. Another advantage of cork is that it has good sound-absorption properties.

✽ IF I GET VINYL TILES, I'LL HAVE ENOUGH MONEY LEFT OVER FOR NEW SHOES!

Sheet vinyl

Today's vinyl is thin, flexible, easy to cut and fit, hygienic, easy to clean and hard-wearing too – the perfect solution to your kitchen and bathroom flooring problems. It is a multilayered sandwich of plastic layers, one of which will probably be cushioned for extra comfort underfoot. The penultimate layer is printed with a multitude of different patterns, designs and colours topped with a clear wear layer.

Vinyl tiles

Imagine sheet vinyl cut up into squares with adhesive on the back. It's as simple as that: just cut to size, peel off the backing and stick to the floor. A much less daunting task than wrestling with a huge piece of sheet vinyl. Vinyl tiles are ideal for a small room with lots of awkward obstacles to negotiate. The basic properties are similar to sheet vinyl: easy to clean, waterproof and durable.

Not just against the wall you can do it on the floor too

CHOOSING FLOOR TILES

36

Tiles are a popular choice of floor covering for use in kitchens, bathrooms and heavy-traffic areas such as hallways. They are especially handy for areas where there are lots of obstacles to negotiate, as they are easy to cut to shape with specialist tools. Ceramic floor tiles are basically the same as wall tiles but are in general thicker, stronger and are designed to be walked on. It is important to note that wall tiles cannot be substituted for use on floors; they will certainly crack under heavy pressure caused by groups of people walking up and down or the weight of large pieces of furniture.

CERAMIC TILES

Ceramic tiles fall into two categories: glazed and unglazed. Glazed floor tiles are not as highly polished as wall tiles, so making them less slippery underfoot. Most are square or rectangular, but there are also lots of lovely shaped ones available, designed to interlock neatly with each other to form patterns. Sizes vary according to the manufacturer, but in general square tiles are 150 mm or 200 mm (6 in or 8 in) square and rectangular ones are 200 mm x 100 mm (8 in x 4 in). This is by no means a hard-and-fast rule – tiles come in all shapes and sizes. Unglazed ones, such as terracotta or quarry tiles, must be sealed

properly to prevent staining. Other materials used for floor tiles include slate and stone, which not only have a natural beauty but also a fabulous texture. Manufacturers, however, do not make life easy for us because they produce a multitude of wonderful styles, patterns and colours to choose from. My advice is have a colour scheme in mind when you go shopping, take a good look at everything and then let your budget be your guide!

TIP
Make sure the shop has more of your tiles in stock – you don't want to run out before you finish the job.

Glazed ceramic tiles

These tiles are generally square or rectangular, but there are many other geometric shapes to choose from. A popular shape is the hexagonal tile, which can be laid with small contrasting coloured square tiles at the intersections to make striking patterns. Choose a monochrome colour scheme or combine two or more colours to make a simple chequerboard design or something more complicated if you feel a touch of creativity coming on.

Quarry tiles

The beautiful warm colours of quarry tiles make them a kitchen favourite. These are usually laid on a mortar bed and must be sealed properly to prevent staining. The most common sizes are 10-cm or 15-cm (4-in or 6-in) squares. Special rectangular tiles with one rounded edge are made to form a neat edge like a skirting board around the tiled floor. Terracotta tiles are quite similar to quarry tiles but are more rustic in appearance. They are handmade and handfired, so the colours can vary from tile to tile, which adds to their individuality and charm.

Not as shiny as wall tiles so you won't slip and fall out of your heels.

Metal tiles

There are also some absolutely fabulous metallic-effect tiles available now for both floors and walls. These are ceramic tiles moulded to look just like metal. Take heed, though; this can be a very expensive option. But go for it if your finances will stretch – very trendy and slick.

TIP

If you decide to do a geometric pattern, try out a few designs on a piece of graph paper first.

TILES ARE EASY TO CLEAN – IDEAL FOR CLUMSY TYPES.

Mosaic tiles

Mosaic tiles are smaller and usually solid in colour. Most mosaics are purchased in square sheets on either a paper or a mesh backing. Use the tiles as they are on the sheet, or apply them individually to form simple geometric patterns. If you feel adventurous, use tile nippers to cut the tiles into smaller pieces to create more elaborate designs. They can also be purchased to form interlocking patterns.

Slate tiles

Real slate or stone tiles can prove to be an expensive option, but they create a lovely effect. Available in a range of natural shades and textures, these tiles are perfect for that rustic look. The texture can be an advantage if the floor is likely to become wet because the surface will be less slippery than a smooth glazed tile would be.

Why worry about the real thing when faking it's so easy

LAYING WOODSTRIP FLOORING

This really is one of the easiest and relatively inexpensive ways to achieve a fabulous wood-effect flooring. Solid wood or laminated planks are glued and slotted together, tongue-and-groove style, to form a beautifully smooth and stylish easy-care floor. The range of wood effects varies from pale beech to darker cherry and mahogany shades. Some manufacturers produce flooring in bright colours and metallic effects for a truly trendy look. The dimensions of woodstrip flooring vary, so measure the room carefully; then read the instructions to calculate the quantity required (*see below*). Remember, this type of flooring requires foam underlay. When you purchase the flooring, buy some suitable underlay at the same time – a heavy-duty one is best to cover any minor imperfections in the subfloor.

TIP

Store the planks in the room in which they will be laid for about 48 hours so they can acclimatize to temperature and humidity.

1 Remove all existing flooring and any nails and screws. Inspect the subfloor carefully to make sure that it is sound and that there are no serious defects. Make any necessary repairs before you begin. Secure any loose boards, hammer in protruding nails and check that all is clean and dry.

2 Cover the entire floor with underlay, joining the strips together with adhesive tape. If the subfloor is made of concrete, lay down a thick sheet of plastic first to prevent moisture rising up through the underlay. If the subfloor is made from uneven timber, you may have to secure a plywood layer to the subfloor first before putting down the underlay. Cut the plywood sheets to fit, then screw them to the timber boards, creating a smooth, even surface.

Estimating the quantities

The packaging will give the average area that the contents will cover. Measure the length and width of your floor space, then multiply the two measurements together to get the floor area. Divide by the figure on the packet to estimate how many packets you will need to complete the job.

3 Take a look at the room and decide which is the longest, straightest wall. This is the best place to start. Open up the flooring installation kit. Usually the kit contains lots of perimeter wedges and a tamping block designed to help you tap the planks home without damaging the 'tongue' side. Begin by laying the first strip with the grooved side facing the wall. Place perimeter wedges next to the wall to create an expansion gap of about 12 mm (½ in).

4 Apply some plank adhesive to the grooved end of the second strip, and insert it next to the first strip. Take a third strip, apply adhesive to the grooved end and then insert it into the second one. Repeat the process until you are close to the other end of the room. Cut the last strip to fit the remaining space, allowing for the 12-mm (½-in) expansion gap of course. Glue as before. You are now ready to begin the second row.

need

* woodstrip flooring
* underlay
* adhesive tape
* flooring installation kit
* plank adhesive
* cloth
* hammer
* wood saw
* pencil
* tape measure
* jigsaw or hacksaw
* strips of quadrant beading
* mitre box
* panel pins

5 Use the offcut of the last strip in the first row to begin the second. This ensures that the joins will be staggered. Continue adding strips and gluing every join along the short and long edges as before. Remember to wipe away excess glue from the surface of the boards immediately using a damp cloth.

6 Place perimeter wedges at the ends of each row to maintain the correct expansion gap all round. Use a hammer and the special tamping block to tap each plank home so it is snug against its neighbour.

✳ GET YOUR OWN PIPE AND SLIPPERS – I'VE GOT A WOODSTRIP FLOOR TO LAY!

7 If you come across any obstacles, such as door architraves or pipes, just cut out the appropriate shape to fit using a jigsaw or small hacksaw. When you have reached the other side of the room and your floor is almost complete, cut the last row of strips neatly to the correct width, allowing for the expansion gap of course. Glue and slot into position. Now you can remove all the perimeter wedges.

8 Cut lengths of quadrant beading to fit around the edge of the room to cover the expansion gap. Use a mitre box to cut the angled joins for the corners (*see page 82*). Nail the beading in place with panel pins.

Time to get down on your hands and knees

LAYING CARPET TILES

38

Carpet tiles are a brilliantly hard-wearing answer to a flooring dilemma in the home. Chosen for their resistance to wear, they are usually thought of as flooring for use in offices and shops. Designed to be loose-laid, these tiles require no underlay or gluing. Carpet tiles are ideal for small areas that receive a hard time, such as hallways or perhaps a home office area or playroom.

need
* carpet tiles
* measuring tape
* sharp craft knife
* paper and pencil
* scissors
* straightedge

TIP

When you buy the tiles, make sure that you get a few spares; they may be handy later to use as replacements in case of stains or burns.

Estimating the quantities

The label on the packaging will give the area that the contents will cover. Calculate the area of the floor by measuring the length and width of the room, then multiply the two. Divide this figure by the figure on the package to calculate how many you will need.

1 The method for laying carpet tiles is the same as for laying vinyl tiles, but they are loose-laid and not glued to the floor. Begin by marking centre lines on the floor. Loose-lay the tiles in a pyramid shape to begin with, using the lines as a guide (*see box opposite*).

2 In general, carpet tiles have a 'nap', that is, the pile runs in one direction. It is therefore possible to make chequerboard patterns or stripes as you lay the tiles. There will be directional arrows on the back of each tile to help you do this. Lay as many complete tiles as you can, then measure and cut the edge tiles to size using a sharp craft knife and a straightedge.

3 If you meet with an awkwardly shaped obstacle such as a door architrave, simply make a paper template of the complete tile and cut it to size with scissors. Trace the shape of the template on the back of the carpet tile.

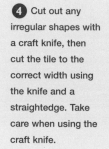

4 Cut out any irregular shapes with a craft knife, then cut the tile to the correct width using the knife and a straightedge. Take care when using the craft knife.

5 Discard the cut-out pieces, then lay the tile into position. Keep any large offcuts or strips of tile as they will be useful for filling in any gaps or awkward shapes you may have left at the end of the job. Also, they may come in handy for repairs later on.

LAYING VINYL TILES

In essence, laying vinyl tiles is the same as laying carpet tiles. Refer to the diagram for positioning. Place the first tile in the angle of the intersecting centre lines.

ARROWS SHOW YOU THE RIGHT DIRECTION. EASY!

1 If the tiles have a specific pattern, they will have arrows on the back to mark this. Make sure that you lay them the correct way. To start with, lay as many complete tiles as you can following the pyramid configuration, as shown below.

2 When you have laid all the full tiles you can, cut the remaining tiles to the correct width to fit around the edges using a sharp craft knife and a straightedge. Make paper templates for any corners, trace the pattern onto the front of the tile, then cut out carefully with a sharp craft knife. Peel off the tile backing paper to expose the adhesive.

TIP

Stack all the tiles in the room for at least 24 hours prior to laying to ensure that they are properly acclimatized to temperature and humidity.

3 Lay the tile into position and press down firmly to make sure that it has properly adhered to the floor.

❋ MMM, THIS VINYL FLOOR FEELS SO GOOD!

ORDER OF WORK

☞ The method for carpet and vinyl tiles is the same. Mark the vertical and horizontal centre of the room with a chalk or pencil line. Align the corner of the first tile to be laid with the centre line, then place the second tile on the other side of the line. Build up as many complete tiles as possible in a pyramid shape to fill half the room, then do the same with the other half. Cut tiles to fill in the edges and awkward shapes last.

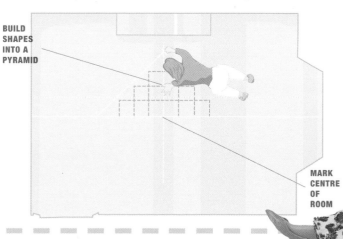

BUILD SHAPES INTO A PYRAMID

MARK CENTRE OF ROOM

For the girl who likes a challenging lay

TILING A FLOOR 40

Ceramic floor tiles are great for kitchens and utility rooms because they are easy to clean – and the terracotta colour is great to look at! The major consideration to be made before you start tiling is the kind of subfloor you have already. Floor tiles can be laid directly over concrete if it is fairly flat, sound and dry. But if you have a suspended floor, i.e. floorboards, it doesn't mean that a tiled floor is out of the question. Ask a surveyor to check if the floor is strong enough to support the tiles; then cover the floor with a layer of 12-mm (½-in) plywood, screwing it securely to the floor. This makes a solid base for the tiles that will not flex. Glazed floor tiles are laid with a waterproof adhesive in a bathroom or kitchen, and quarry tiles are laid in a bed of mortar. Both types should be laid with a flexible adhesive if the subfloor is suspended.

1 To begin, remove all existing floor coverings and then prepare the floor as necessary. Mark out the floor as shown in the diagram on page 67. The method of setting out floor tiles is similar to that of laying carpet and vinyl tiles – start from the centre lines and work out the spacing so that you are left with reasonably wide border tiles. The only difference is that you must begin laying out the tiles in the corner furthest from the door.

TIP

As quarry tiles are thicker than glazed ceramic floor tiles, care should be taken when cutting to size. Allow plenty of spare tiles in case it takes a few attempts!

Start tiling from the corner that is furthest from the door.

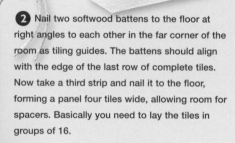

2 Nail two softwood battens to the floor at right angles to each other in the far corner of the room as tiling guides. The battens should align with the edge of the last row of complete tiles. Now take a third strip and nail it to the floor, forming a panel four tiles wide, allowing room for spacers. Basically you need to lay the tiles in groups of 16.

3 Mix up the mortar in a bucket as per the manufacturer's instructions. Using the trowel, lay a thick bed of mortar between the battens, then begin to lay out the tiles with spacers between them. Mix the mortar in small usable quantities as the job progresses. Don't try making a huge bucketful – it'll be really heavy and you don't want to put your back out!

Estimating the quantities

Floor tiles are generally sold in boxes, but can be bought individually sometimes. The manufacturer will give the area that the contents will cover. Work out the area that needs to be tiled, then use the manufacturer's notes as a guide to the quantity you'll need. Bear in mind that you will make mistakes, so allow for between 5 and 10% extra.

need
* tiles
* battens
* mortar
* bucket
* trowel
* spacers
* spirit level
* heavy-duty tile cutter

*I'VE ALWAYS WANTED TO LAY IT ON THICK WITH A TROWEL.

4 Trowel on more mortar as you go. Do not try to rush a tiling job. It's worth taking your time to make sure that all the spacings are correct. By the way, the plastic spacers are specially designed to be left in the mortar, so you don't have to try and pick them all out!

5 It is very important that the tiles are level, so check and double-check as you go, tamping each row of tiles down gently using an offcut of batten and the handle of a hammer. Then double-check again. When the tiles have been well tamped down into the mortar, use a spirit level just to make sure that they truly are level. Little irregularities will be really noticeable when the floor is complete. Remove the battens once each group of 16 tiles has been completed.

6 Continue working along the battens in groups of 16 until the floor is covered with all the whole tiles (this is called laying 'field' tiles). Plan on finishing at the door side of the room so you can retreat to the sofa for a nice rest without having to walk all over the freshly laid tiles.

7 Wait at least 24 hours before walking on the floor, either to fill in with mortar or to lay the border tiles. Use the tile cutter to cut border tiles to the correct width to fit the gap left around the edge. Lay in mortar as before, then fill in the spaces with a drier mortar mix, brushing away the excess with a wire brush.

Celebrate with a cup of tea and a biscuit (choc chip of course!) - you deserve it.

How to get to grips with a big roll of vinyl with the minimum of fuss

LAYING A VINYL FLOOR

41

Laying sheet vinyl in a fairly small room such as a bathroom or kitchen should present no problem at all. However, if the floor space is large, you may want to ask a friend to help you. In general, most rooms have one reasonably straight wall. Take this as your starting point; then do all the fiddly bits later. You are likely to come across lots of those if you are laying it in the bathroom!

need
* vinyl
* offcut of batten
* nail
* hammer
* cold chisel
* heavy-duty scissors
* sharp craft knife

* I KNOW! WITH THE MONEY WE'VE SAVED I CAN BUY SOME NEW GOLF CLUBS!

TIP
Leave the vinyl in the room for at least 24–48 hours prior to laying, so it becomes properly acclimatized and therefore less likely to warp.

1 Take your offcut of batten and hammer a nail through it about 5 cm (2 in) from one end so that the point just protrudes on the underside. This is your scribing tool. It will help you cut the correct shape in the first side of the vinyl. You may think that the straight-looking wall is in fact straight, but the truth is it probably isn't, so use the scribing tool to get the right fit.

2 Lay the sheet parallel to the first wall, about 3.5 cm (1½ in) away from the wall. Place the end of the scribe against the wall and drag it along the length of vinyl. The point of the nail will mark a cutting line on the vinyl to ensure a perfect fit (there will of course be an overlap at each end). Use heavy-duty scissors to make this first cut. Place the vinyl against the wall again and make any necessary adjustments.

3 The next step is to cut a square or triangular notch (or a release cut) at each corner so that the sheet lies fairly flat. Press the vinyl firmly into the angle made by the floor and wall. You can use a cold chisel to help you do this.

4 Then use your craft knife to cut along the crease carefully. Take your time with this, because vinyl can be tough to cut and it's easy to exert a little too much force and make a slip.

5 You will undoubtedly come across shaped obstacles such as a basin pedestal or toilet base. Fold the vinyl back from the wall, then press the sheet to the base of the obstacle. Make one cut running from just above floor level upwards towards the edge of the vinyl.

6 Draw the blade through the vinyl from the base of the obstacle towards the edge, making triangular-shaped 'tongues'. Work around the curve in this way. The vinyl should now follow the shape of the obstacle. Cut away any excess, but don't be too ruthless as the vinyl has to fit around the back of the pedestal or base too.

7 When you are happy with the approximate fit, trim close to the base with the knife. Cut the vinyl to fit any other odd shape in the same way. Templates can be cut and followed to fit neatly around door frames.

Wear protective gloves if you aren't used to using a craft knife.

Lay, lady, lay... a lovely new carpet in an afternoon

LAYING CARPET

Laying a small- to average-sized piece of carpet isn't too hard, as long as you can lift the roll of carpet easily. If things get tough, ask some friends to help. For large rooms, however, it's probably better to get the carpet fitters in. If your budget is tight, check out your local carpet stores and warehouses, as they always have inexpensive remnants and small pieces of carpet available. In fact they often have long leftover strips a few feet wide – ideal for a small staircase or hallway. The next few steps show you how to fit both foam-backed and hessian-backed carpets.

TIP
If you need to join two pieces of carpet to make the correct size, do so using the double-sided tape – and don't forget to match any patterns!

need
* carpet
* double-sided floor-to-carpet tape
* Stanley knife

FOAM-BACKED CARPET

1 Remove old carpet and underlay. Make sure the subfloor is clean and dust-free. Most foam-backed carpets do not require underlay, which is great if you are short of cash and time, but remember, this type of carpet is best for light domestic use, in an area not likely to receive rough treatment.

CHOOSE YOUR CARPET CAREFULLY – SOME CAN GIVE YOU NASTY CARPET BURNS!

2 First, fix a wide strip of double-sided floor-to-carpet tape around the perimeter of the room. This will hold the carpet in place and prevent it from shifting.

3 It's best to begin at the longest, straightest wall, before dealing with any awkward obstacles. Peel off the paper backing strip of the floor tape to reveal the adhesive.

4 Lay the carpet onto the adhesive tape. Use the same procedure as for sheet vinyl to fit the rest of the carpet into corners and around obstacles (see page 71).

✳ GET RID OF THAT PREHISTORIC CARPET NOW!

HESSIAN-BACKED CARPET

1 Hessian-backed carpet requires foam underlay and edge-fixing strips called gripper rods. These are thin wooden battens with sharp teeth that are designed to hold the carpet firmly around the perimeter of the room. They should always be handled with care.

2 Place the gripper rods around the perimeter of the room. They come with nails attached – all you have to do is hammer them in. Cut them to size using a hacksaw. Make sure that the teeth point towards the wall in order to grip the carpet properly. Nail them in place allowing about a 1-cm (⅜-in) gap between the gripper and the wall.

3 Cut the underlay to size so it butts neatly against the gripper rods. Join pieces together with double-sided tape if necessary. Then staple to the floor at intervals around the room's perimeter.

TIP

Refer to the previous sections on fitting carpet and vinyl tiles and sheet vinyl for tips on how to cut around obstacles, make templates and so on – the principle is basically the same for carpets.

4 Roll out the carpet and use your Stanley knife to trim it roughly to size. Begin fitting at the longest wall and the wall adjacent to it.

5 Use the bolster chisel to push the carpet onto the gripper rods so the teeth hold it firmly, then tuck the edge of the carpet under the skirting board. It is important that this first edge is held securely on the teeth of the gripper rods. You will need to stretch the rest of the carpet so that it lies flat with no creases, and you don't want the carpet pulling off the gripper rods mid-stretch!

6 Stretch the carpet towards the other two corners of the room. Then follow the same procedure, hooking onto the gripper rods, trimming and then finally tucking the edge under the skirting board.

need

* carpet and underlay
* hammer
* hacksaw
* gripper rods
* staple gun
* bolster chisel

Onwards and upwards — *tackling the stairs*

LAYING A STAIR RUNNER

43

Use the same principles as for laying hessian-backed carpets. You need gripper rods to hold the carpet firmly to the steps and to prevent slipping when you walk up and down the stairs. Another point to remember – on a stair carpet the pile should run downwards.

Kneel on a cushion while working to protect those delicate knees.

1 Using a small hacksaw, cut the gripper rods to fit the stair width – you'll need two grippers for each step. Hammer one to the back of each tread (horizontal part of the step) and at the base of each riser (vertical part of the step). When hammering the gripper to the riser, make sure that you allow a gap of about 15–18 mm (⅝ in–¾ in) between it and the first tread gripper. This creates space for the carpet to be pushed into, using the bolster chisel. The teeth of the gripper rods will hold it securely in place in the angle of each riser.

2 Cut individual pieces of underlay to fit over each tread and the riser below so they butt up against the gripper rods, but not over them. Lay in position and either staple in place or use small tacks. Cover the entire staircase in this way. Use your bolster chisel to press the carpet onto the gripper rods, making sure that the carpet is held securely onto each tread and riser.

CARPETING WINDING STAIRS

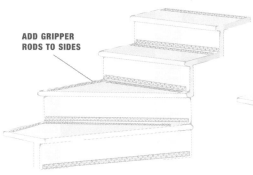

ADD GRIPPER RODS TO SIDES

FOLD CARPET TO FIT STAIRS

☞ To carpet a winding staircase edge to edge, treat each step separately. Attach gripper rods as for straight stairs, then add lengths of gripper to the side of each tread. Working from top to bottom, cut pieces of carpet to fit each step in turn, ensuring the 'nap' of the carpet lies in the same direction for each.

☞ To carpet your winding stairs with a stair runner, do not fix gripper rods to the risers of each step. Working from the bottom to the top, fold the carpet as shown above so that it fits flat against each step. Hold each section in place with 4-cm (1⅝-in) long tacks.

need

* carpet and underlay
* gripper rods
* hacksaw
* hammer
* Stanley knife
* bolster chisel
* stapler or tacks
* scissors

THIS IS EASIER THAN IT LOOKS AND SOOO IMPRESSIVE!

LAYING CARPET OVER A BULLNOSED (CURVED) TREAD

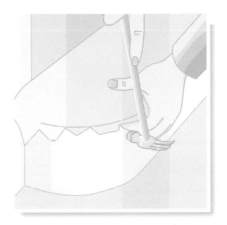

1 Lay carpet across the top of the tread, then use scissors to cut the edge roughly to the shape of the curve, allowing about 5 cm (2 in) excess all around. Cut small wedge-shaped portions from the edge of the carpet so that it will fit neatly around the bullnose part of the step.

2 Press the triangular tabs around the bullnose using your thumb or fingers to make sure it fits neatly. Trim away any excess so that the carpet lies flat. When you are satisfied with the fit of the carpet, nail the carpet to the riser underneath the bullnose, using carpet tacks.

3 Cut a strip of carpet to fit exactly around the riser underneath the bullnose. Place the strip in position, then hammer tacks along the upper and side edges. Do not tack along the lower edge – the tacks will probably be visible and unsightly.

3 Cut the carpet to fit the width of the stairs, then begin fitting from the top towards the bottom. Use the bolster chisel to tuck the top edge of the carpet onto the gripper rod at the back of the tread at the top of the stairs. Work downwards over each tread and riser, using the chisel to push the carpet onto the gripper rod at each angle.

4 Fit gripper rods, underlay and carpet to the landing. Overlap the end of the carpet over the top step and down the riser below.

❋ HOWDY – YOU CAN HANDLE MY RISER ANY TIME!

And for all you creative girls out there

DON'T LIKE IT? YOU CAN EASILY SCRAPE IT OFF AND START AGAIN!

DECORATIVE FLOORING IDEAS

44

Well, we've covered the very basic elements of carpeting, tiling, laying vinyl and painting but all this is pretty straightforward. If you feel a rush of creativity coming on, however, there's definitely a lot of scope for doing it on the floor! Think colour, think patterns, think borders, think stripes, think bold motifs – just about anything is possible. Take a sketchbook and pencil and do some creative drawing to see what you can design. Alternatively, browse through a pile of style magazines for inspiring ideas and colour schemes.

Stencilling

Stencils can really transform a dull room. Try a simple geometric border or an elaborate stencilled pattern in the corner of a room. Mark out the design carefully on the floor, then set to work with your paintbrush or small roller. You can use emulsion paint, or one of the wide range of paints designed for stencilling and stamping projects.

SCALE DRAWINGS

One final point that I feel I must stress – scale drawings. If you're creating a design on the floor (unless it's meant to be random and arty), you'll need to follow a template or pattern. Buy some large sheets of graph paper, measure the room carefully and then draw the outline on the graph paper. You can then sketch the desired design or pattern, translate it directly to the floor or piece of flooring and cut or paint to the right size. Success depends on precise and accurate measuring, so be careful.

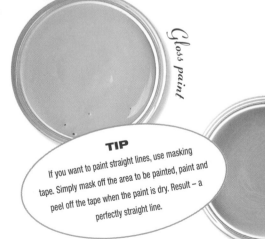

Gloss paint

Metallic-effect paint

Floor paint

TIP

If you want to paint straight lines, use masking tape. Simply mask off the area to be painted, paint and peel off the tape when the paint is dry. Result – a perfectly straight line.

✳ ALL I NEED IS SILVER-PAINTED FLOORBOARDS AND I'VE GOT MY OWN DISCO!

'Trompe-l'oeil' rugs and runners

'Trompe-l'oeil' means 'deceive the eye', a decorative effect designed to trick you into thinking that something is there when it isn't. This idea can be used to create fake painted rugs or runners on wooden floors in hallways, on stairs or in front of a hearth. Choose simple geometric designs, or if you feel ambitious, choose something with a more complicated pattern. For an authentic look, try painting tassels at both ends – or a cat and a pair of slippers on a hearth rug!

Painted floorboards

Painted floorboards are easy to clean and repaint. You don't have to restrict yourself to one colour either. Why not create a retro fifties-style striped living area: beige, brown, pale blue, then shocks of hot pink and vibrant tangerine. If you're not feeling so ambitious, try a simple two-tone chequerboard effect to simulate tiles.

MORE FABULOUS IDEAS

Painted concrete Concrete may not be very inspiring, but if you've taken up an old threadbare carpet and found concrete underneath, this is a great quick fix. The choice of colour is fairly limited, but anything is better than the basic concrete colour. Give it a try. If you need to lay a new concrete floor, perhaps in a basement or utility room, it is possible to buy concrete colouring pigments.

Painted vinyl Manufacturers have come up with this fantastic product to paint onto sheet vinyl flooring and vinyl tiles. This is a superb quick fix if your vinyl flooring is in good condition but the colour or pattern makes you cringe. Simply prepare the floor as per the manufacturer's instructions, then apply the vinyl paint. Use one colour all over, or be creative with bold motifs or patterns.

Carpet borders Most carpet stores sell borders to complement their plain carpets – these can be plain contrasts or stripes or maybe a pattern. Try including a border when you lay a carpet to create a bit of interest around the edge. The pieces are stuck together with strong carpet tape on the back. This is pretty easy to do, it's the same technique that you use to join pieces of carpet to cover a large space.

NAILING IT

The art of construction: putting things up, banging them together, nailing them in and a few original projects too.

Basic construction

YOU JUST BANG THINGS IN, DON'T YOU?

IN THIS SECTION we look at a few simple construction projects that you can do around your house or flat, such as putting up shelves and curtain rails and how to assemble flatpack furniture and make new tongue-and-groove bath panels. And to finish off, there are some slightly more complex projects – these will provide a good opportunity for you to get creative and adventurous.

USING CONSTRUCTION TOOLS OUCH!

HAMMERING

When hammering a nail into a piece of wood, take the nail between thumb and forefinger and position it on the wood. Tap a few times gently with the hammer, so that the nail just pierces the wood and can stand up by itself, then strike the head of the nail a few times to drive it right into the wood, keeping the hammer square to the nail head. If you bend the nail while hammering, pull it out and start again with a new nail.

SCREWING

The most important thing to remember is to use the correct screwdriver to fit the screw: slot-head or cross-head. Next, do you know which way you're meant to be screwing – clockwise or anticlockwise? When you're trying to remove a really tight or painted-in screw, it's easy to forget which way you are supposed to be turning. Just keep this in mind: 'righty-tighty, lefty-loosey'. Simple.

DRILLING

A power drill is probably one of the first power tools that you will buy and it will be the one that you use most frequently. Basically, a power drill has adjustable jaws into which you can fit all sorts of drill bits, used for making holes of all sizes. When you use a power drill, make sure that it is square with the surface that you are drilling into and that you have a firm grip of it before you switch it on. A variable-speed drill is useful because you can start slowly, then increase the speed as you go or choose different speeds for different jobs.

JIGSAWING

A jigsaw is a really useful item to have in your tool kit. It is generally used for making curved cuts in sheet material, although both straight and angled cuts can also be made. Simply rest the sole plate of the jigsaw on the material to be cut, switch the saw on, then gently move the cutting blade through the material along a pre-marked cutting line. Take care not to force the blade at all or twist it in

SAFETY

☞ Always wear safety goggles in case a small piece of something flies off during drilling or sawing and hits you in the eye.

☞ When sawing timber, MDF or anything that might create dust, wear a dust mask.

☞ Unplug the tool before changing any blades or bits, etc., then make sure that the attachment is secure before attempting to use the tool again.

☞ If you are using extension leads for your tools, make sure that the flex is fully uncoiled; this will prevent the flex from overheating.

☞ Make sure that plugs and flexes are properly attached and safe. Never pick up a power tool by its flex.

any way. Always switch it off first, let the blade stop and then remove it from your work when finished. Change the blades frequently – using a dull blade is hard work and can damage the material you're cutting.

Mirror, mirror...

HANGING A MIRROR OR HEAVY PICTURE 45

To hang a small picture, a piece of string stapled to the back may well suffice. For a large picture or a heavy mirror with a chunky frame, a different approach is needed.

need

* plate hangers and screws
* pencil
* bradawl
* spirit level
* drill with appropriate bit
* wall plug and screw (for plasterboard or masonry)
* screwdriver

Flat mirror plates

1 Lay the frame face down on your work surface, and mark the positions of the hangers about 10 cm (4 in) from the top and bottom edges of both sides.

Wall plugs

2 Place the brass picture hangers in position. Then make small pilot holes using a bradawl to correspond with each screw hole; this will make it easier to drive in the screws. Screw each hanger firmly to the frame. For a large frame use two hangers on each side, but for a smaller frame a hanger halfway down each side will suffice.

3 Choose the hanging position and mark the screw holes on the wall with a pencil. Use a spirit level to make sure that the markings are horizontal. Insert a drill bit in the electric drill to match the wall plug and screw. Drill holes in the marked positions. Use your thumb to push or a hammer to tap a wall plug into each of the holes.

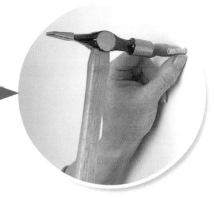

4 Screw the mirror or picture securely into position. If the frame is very heavy, you may need a friend to help you hold it steady while you drive the screws home.

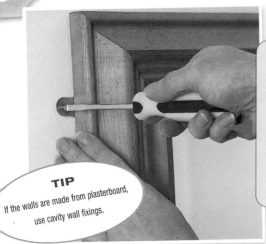

TIP
If the walls are made from plasterboard, use cavity wall fixings.

PICTURE RAILS

If you want to hang pictures from a picture rail, you must reinforce the rail first – this is so that your picture doesn't fall off the wall and take the rail with it. Make a mark on the rail about every 1 m (3 ft), then drill and countersink a screw hole. Insert a wall plug and then drive in the screw. Fill the countersunk hole afterwards with wood filler.

Putting up decorative wall accessories

PUTTING UP A DADO RAIL

46

Dado rails, picture rails and decorative coving add a special touch to your decor, especially if you have a modern flat or house where these features are rarely included. Dados and picture rails are usually made from moulded wood, whereas coving is generally made from plaster or, in some cases, polystyrene.

need

* dado rail
* pencil
* spirit level
* instant grip adhesive
* mitre saw

CUTTING MITRES

☞ A mitre is the join between two pieces of wood where the ends are cut at an angle – useful for decorative mouldings. For internal corners, cut the moulding to fit square to the corner, then mark the angle away from the corner. For external corners, cut the moulding so it protrudes beyond the corner; mark the corner position on the moulding. Cut the angle from the pencil mark away from the corner. Set your mitre saw at an angle of 45 degrees.

---

1 Using a pencil and spirit level, mark the position of the dado rail, usually about 1 m (3 ft) from the top of the skirting boards. This is only a general rule; if you want to put it a little higher or lower, it's up to you. Cut the moulding to fit around the room using a mitre saw. Cut 45-degree angles at the ends so they fit together neatly around external and internal corners. Remember to mark the direction of the mitre carefully before cutting because it's easy to get internal and external angles confused.

2 Apply instant grip adhesive liberally in a wavy line all the way along the back of each dado strip. The adhesive will be strong enough to hold the dado rail in place without the aid of screws.

3 Press the dado against the wall firmly, aligning its top edge with the pencil line. If you need to join strips to achieve the correct length, simply glue and butt-join to another strip.

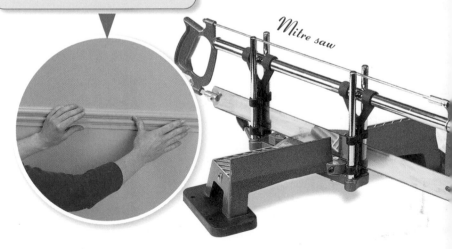

Mitre saw

PUTTING UP COVING 47

Coving is a great way to finish off a room and also has the benefit of covering up the unsightly wobbly line that occurs between the paint on the wall and the ceiling.

GEE, THAT CEILING NEEDS COVING!

1 Hold the coving at the angle between wall and ceiling, then trace lightly with a pencil along the top and bottom edges. Using the point of a small trowel, score diagonally to make a crosshatching pattern along the wall between the pencil lines and the angle of the ceiling. This is to create a good key for the adhesive.

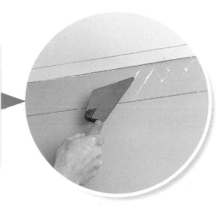

2 Cut the coving strips at a 45-degree angle at the ends to fit neatly around internal and external corners. You can join pieces together by gluing and butt-joining the ends. Cover gaps with wood filler, then sand smooth when dry.

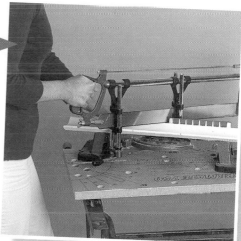

3 When the pieces have been cut to size, mix the coving adhesive according to the manufacturer's instructions, and apply it to the back of the coving using a scraper or spatula. Carefully press the coving into position. There will be marks on the back of each piece indicating which is the wall edge and which is the ceiling edge.

need

* coving
* mitre saw
* pencil
* trowel
* instant grip adhesive
* coving adhesive
* scraper or spatula

PUTTING UP A PICTURE RAIL 48

☞ If the picture rail is for decoration only, then fix it to the wall in the same way as for the dado rail. In general the rails are placed about 50 cm (20 in) down from the ceiling.

☞ If the rail is for hanging pictures, drill and countersink a screw hole every 1 m (3 ft) along the rail, then glue in place. Drill another hole through the first into the wall, plug, then drive in a screw. Fill the countersunk hole with wood filler and paint over afterwards.

Flatpacks — a girl's guide to self-assembly

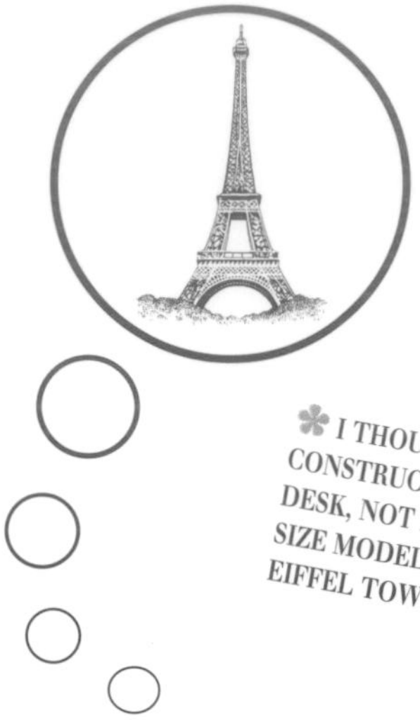

THE ULTIMATE GUIDE TO FLATPACK SUCCESS

I'm sure most of you have spent many frustrating hours struggling with so-called 'easy-to-assemble' furniture kits. Yes, they can be confusing, and there are lots of bits and pieces that look like nothing you've ever seen before, but there's no need to panic. Flatpacks these days are much simpler to understand than they used to be. Over the years manufacturers have done a lot of market research and discovered that we like it plain and simple. Most instructions have lots of diagrams and very little text, so you just have to look at what you've got and match it with what you see in the diagrams.

WHAT'S IN THE BOX?

So you've hauled your flatpack home and opened up the box. Inside there will usually be lots of polystyrene packaging and, more importantly, there'll be pieces of MDF or coated chipboard with pre-drilled holes and slots: possibly doors and drawer units, together with numerous bits and pieces that appear to make no sense at all. There will also be a small bag full of nuts and bolts or screws and strange-looking items, possibly an allen key, wheels, drawer runners, wooden dowels and a container of glue. And last, but not least, you should have a set of instructions! Do not lose this vital piece of paper – it is extremely important and will show you what everything is and what you should do with it.

* I THOUGHT I WAS CONSTRUCTING A DESK, NOT A LIFE-SIZE MODEL OF THE EIFFEL TOWER!

THE PERFECT DESK KIT

49

Empty the contents of the small bag onto the floor or, better, onto a tray so that you don't lose any pieces, and sort them into groups of similar objects. Each will be illustrated in the instructions, along with a note of how many there should be. Now move on to the big parts. Lay everything out exactly as it is in the instructions, so you can compare. Similar-looking pieces may have different pre-drilled holes, for example, so if you use the wrong piece, the item won't fit together.

1 This is what you're likely to see in the small bag of bits and pieces: dowels, screws, plastic screw caps, nails, screw fixings, etc. Check them against the list in the instructions. Try not to lose anything at this stage – it may be crucial later on!

TIP

Get yourself a rubber mallet. This tool is useful for driving home dowel joins without damaging the surface material. In flatpack land you just never know when you might need to hit something without leaving a mark!

need

* hammer
* screwdriver
* rubber mallet
* possibly glue if not provided

2 Most flatpacks have dowel joins. The dowels themselves look like little ridged rods of wood, and the flatpack pieces to be joined together will have pre-drilled holes to fit the dowels. All you have to do is squeeze a little wood glue into each of the holes on one piece to be joined, then use a hammer to tap the dowel about halfway into each of them. Locate the corresponding piece, place glue in the holes as before, then slot the first dowels into the holes of the second piece. Finally, use the rubber hammer to tap the dowels in, making sure the pieces fit neatly together.

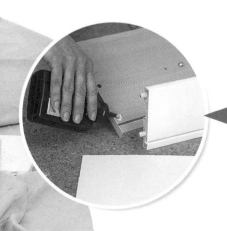

3 Most flatpack drawer systems are the same: three sides of the drawer are joined together or are sometimes joined with plastic hinges that can be folded to the correct shape. Each side has a slot near the lower edge, where a piece of hardboard slots in to form the drawer base. The drawer front is then screwed or dowelled to the drawer body. A similar method is used for cabinets and bookcases.

4 Now it's time for the rubber hammer treatment. Finish off your furniture item with a quick going-over with the rubber hammer. This will make sure that all the joins are secure and firm.

The art of putting up shelves

BUILDING A SHELF INTO AN ALCOVE 50

Alcoves are wonderful little niches for putting shelves into, providing excellent areas for books, ornaments or audio/visual equipment. The simplest way to do this is to use three wooden battens, one to fit across the back of the alcove and one to fit each side. The shelf sits on top of the battens, providing a sturdy surface on which to put your items.

1 Cut a batten to fit across each side of the alcove. Trim one end to a 45-degree angle (this makes it neat at the front), then cut a length to fit across the back of the alcove minus the width of the two side battens. Drill and countersink a clearance hole through the batten about one third and two thirds of the way across. Hold the batten up to the side of the alcove at the correct height, and use a spirit level to make sure it's straight. Draw a pencil line along the top of the batten to mark the exact position, then mark the drill holes with a bradawl.

2 Drill screw holes into the wall at the points previously marked with the bradawl. Remember to hold the drill square to the wall, so the screw hole will be straight. Tap wall plugs into each hole, then screw the batten in place. A power screwdriver is a very useful tool if you have lots of screws to insert – they make life a lot easier.

3 Cut the shelf to fit (*see right*). Balance one end of the shelf on the fixed batten, and supporting it with one hand, lower it into position. Place the spirit level on top to check that it's straight, then draw a pencil line along the underside of the shelf, along the back and opposite side edge. This shows the positions for the remaining two battens. Attach the battens to the wall in the same way as the first. Then rest the shelf on top. If you like, you can screw the shelf to the batten to make it more secure.

Even a shelf needs help to find its own level.

need

* timber for shelf (or MDF)
* lengths of 2.5-cm x 2.5-cm (1-in x 1-in) softwood battens
* wood saw
* measuring tape
* spirit level
* drill with appropriate bit, countersink bit
* bradawl
* screws and wall plugs
* sliding bevel
* jigsaw

Sliding bevel

BASIC

HOW TO CUT A SHELF SO THAT IT FITS PROPERLY!

☞ A sliding bevel is a really useful item to have when putting up shelves. It will ensure your shelves fit perfectly every time. Before you cut a shelf, measure the width at the back of the alcove itself, then across the front – more often than not there will be quite a difference between the two measurements. This suggests that the edges of the shelf must be cut at an angle to fit properly.

This is where the sliding bevel comes in. Open up the blade, then rest the wooden handle along the wall at the back of the alcove in one corner. Push the blade so that it rests on the batten; this will give you the exact angle to cut. Mark the centre of the shelf. Then align the midpoint of the back measurement on the centre line. Now place the bevel on the shelf at this point and mark the angle of the cutting line exactly. Do exactly the same for the other corner.

CUTTING A SHELF TO FIT

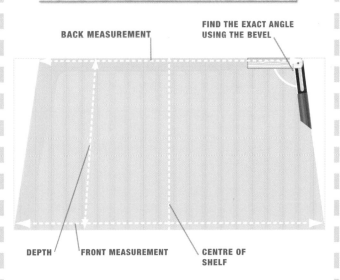

BACK MEASUREMENT

FIND THE EXACT ANGLE USING THE BEVEL

DEPTH **FRONT MEASUREMENT** **CENTRE OF SHELF**

USING DECORATIVE WOOD BRACKETS 51

☞ There are two ways to fix shelves with brackets – screw the brackets to the wall first or attach them to the shelf first. In the case of these chunky decorative wooden brackets, you need to decide the height of the shelf, place the bracket against the wall, then mark the drill holes with a bradawl. The bracket will have pre-drilled holes already. Drill and insert a wall plug into the wall, then screw the bracket in place. Use the spirit level to mark the position of the second bracket a suitable distance from the first – long enough for the shelf to sit across both brackets comfortably. Screw the other bracket in place, then lay the shelf on top.

Decorative brackets

need
* spirit level
* pencil
* drill with bit
* screws and wall plugs
* screwdriver
* bradawl
* shelf cut to size

Hey... more shelving

USING SHELVING SYSTEMS AND ADJUSTABLE SHELVING

52

Most DIY stores sell shelving systems. Made of metal, the systems consist of upright bars in various lengths, with slots along the length. Metal struts fit into the slots, forming supports for the shelves. This form of wall storage is useful because the struts can be moved to create shelf space of different sizes. If the walls are really uneven, the uprights may not be able to lie flush to the wall, so choose another form of shelf support in this case.

AND

USING METAL BRACKETS, FIXED SHELVING

☛ You can also screw metal brackets to the shelf first and then attach the brackets to the wall.

☛ Lay the shelf face downwards on a flat surface, then position the brackets close to the back edge. Screw securely in place.

☛ Now take the shelf and place it on the wall at the correct height, then place the spirit level on the shelf in order to check that it is straight.

☛ Mark the screw-hole positions throughout the brackets using a bradawl. Drill and plug the screw holes, then screw the brackets with shelf attached to the wall.

need
* shelving uprights
* shelving struts
* bradawl
* pencil
* spirit level
* drill and drill bit
* screws and wall plugs

1 Decide on the position of the first upright. Mark, drill and plug the top screw hole, then loosely screw the upright in place. Check that it is vertical using a spirit level.

2 Push the upright to one side, pivoting on the top screw, then drill and insert wall plugs into the other screw holes. Allow the upright to swing back into the vertical position, and drive in the screws. You can now tighten up the top screw. Rest the end of the spirit level on the top of the first upright, and mark the position of the top of the second upright. It is important to check the horizontal as well as the vertical to ensure that the shelves will be level.

3 Place the second upright in position, then mark the top screw hole. Drill and plug the hole, then loosely fix the upright in place. Check the vertical again using the spirit level, then mark, drill and plug the other screw holes as before. Drive in the screws to hold the upright firmly in position. Slot in the adjustable metal struts at the desired positions and rest the shelves on top.

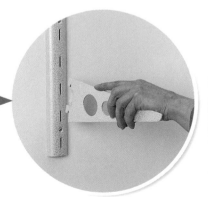

Easy ways to jazz up your kitchen

CHANGING DOORS ON CABINETS 53

Have you just moved into a new flat or house and hate your kitchen? No cash to get a new one? You don't have to tear out the old kitchen and install a new one right away if your finances are a bit delicate, just give it a quick-fix treatment.

You can transform the look of your kitchen just by taking off the old doors and replacing them with new ones. Most kitchen cupboard carcasses are one of a few standard sizes, and new ones will have holes pre-drilled for hinges. New cupboard doors are readily available from DIY stores, and you can always customize them with paint or new handles.

PUTTING UP A SMALL CABINET 54

This fairly dull-looking gadget is a battery-operated stud finder – studs are vertical pieces of wood that you find inside plasterboard walls. If you're going to hang a cabinet on a masonry wall, there is no problem; heavy duty screws and wall plugs will do the job. However, plasterboard is a little different, because the wall has a cavity inside. That in itself should not present a problem if the fixture is fairly lightweight – you can use a suitable cavity wall fixing. On the other hand, if the fixture is to be a load-bearing one, such as a cupboard or a bookshelf, then you need to screw directly into the studs or there's a risk of your fixture coming loose. Use the stud finder to locate the studs; then hang the cabinet securely.

GIVE YOUR KITCHEN AN INEXPENSIVE REVAMPING BY PAINTING THE CUPBOARD DOORS. THIS PICTURE SHOWS A HAND-PAINTED TORTOISESHELL FINISH IN A TRADITIONAL KITCHEN.

Stud finder

FIND ME A STUD, BIG BOY!

USING A STUD FINDER

1 A battery stud finder uses electronic signals to locate studs or joists through dry wall material. To begin you must calibrate the unit. Hold the unit against the surface to be checked, make a firm contact and press and hold down the activation switch. The red light will flash, then the green 'ready' light will go on and stay constant. Continue to hold the activation switch down throughout all the following procedures.

2 Slide the unit across the surface. When you approach a stud or joist, the red light will flash, then remain constant when it detects the edge. Mark this position, then double-check by repeating the process from the other direction.

3 The stud finder will identify pipes or electrical wiring as a stud. Usually, joists and studs are regularly spaced. If you find anything in between, it is wise to assume that it is wiring or a pipe of some sort. If in doubt, do not drill into it.

Learn how to deal with poles, or it's curtains for you...

ONE FALSE MOVE AND THOSE CHINTZ CURTAINS ARE HISTORY.

CURTAINS ETC.

Curtains make a house a home, and whether you want to change the curtain track you have already and replace it with a pole or vice versa, it's quite a simple operation. Once again the choice is enormous, from simple styles to incredibly ornate poles and finials. In general, curtain track is metal or plastic and can be fixed to the wall or suspended from the ceiling. Most can be bent or curved slightly to fit around bay windows or corners.

POLES AND TRACKS

Curtain poles are made from wood or metal and have a decorative finial at each end. It is possible to have a pole fit a corner, but normally they are used for straight runs only. The general rule of thumb when measuring for a rail or pole is to measure the width of the window, then add at least 15 cm (6 in) at each side. This allows space for the curtains to hang neatly and not cut out any light when they are opened up. A few words of guidance concerning positioning of the track and pole: usually the hardware is positioned at least 15 cm (6 in) above the top of the window. If the window has a reinforced concrete or galvanized beam across the top, behind the plaster, it may be impossible to drill a hole in it. The answer is to attach the track or pole a little higher than the reinforcement. All tracks and poles come in kit form with full instructions, complete with all fixings, screws and wall plugs needed. Read through carefully before you begin, to familiarize yourself with the procedure. Most kits are available in a range of standard lengths. Choose one closest to the length you require, then trim to fit using a hacksaw.

PUTTING UP A CURTAIN RAIL

55

need

* curtain track kit
* spirit level
* pencil
* drill with bit
* wall plugs
* screwdriver
* bradawl

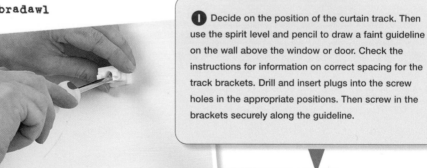

1 Decide on the position of the curtain track. Then use the spirit level and pencil to draw a faint guideline on the wall above the window or door. Check the instructions for information on correct spacing for the track brackets. Drill and insert plugs into the screw holes in the appropriate positions. Then screw in the brackets securely along the guideline.

2 Check that you're screwing the brackets in the right way up; you may come unstuck at the next stage if the slots aren't in the right place.

PUTTING UP A CURTAIN POLE

You can usually attach pole brackets to the wall with wall plugs and screws, but in this case the window had a handy wooden frame. I just screwed the brackets directly into the frame, which was helpful because there wasn't enough room between the frame and the coving to fit the bracket posts.

1 Hold the plastic bracket post in position on the frame; then mark the screw hole positions using a bradawl. Drill pilot holes at each mark.

need

* **curtain pole kit**
* **bradawl**
* **drill with bit**
* **screwdriver**
* **spirit level**

2 Screw the bracket to the wooden frame, then repeat at the other side of the window. It is worth using the spirit level now, just to check that the brackets are level and horizontal.

TIP
Some older properties may have wooden-framed windows so it is just a matter of screwing the brackets into the wooden frame.

3 Slot the track-locking clips into each bracket. These little devices hold the track firm to the brackets. Fit the track onto the brackets, checking that the track extends an equal amount at each side, then push each of the locking clips to a horizontal position. This will lock the track in place.

4 Slip the sliders onto the track; then insert an end stop at both ends of the track. You can now hang your curtains.

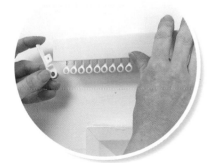

3 Fit the bracket extension onto the plastic post and then insert and tighten the grub screw at the base. Slip all the curtain rings onto the pole, and then rest the pole on the brackets. Check that the pole extends an equal amount on each side, then insert and tighten the grub screw that holds the pole to the bracket. Finally, place a finial at each end of the pole.

Keeping it covered — pelmets have never been so much fun

GIRL-TASTIC!

PUTTING UP A PELMET 57

When the curtains are up, you may decide to add a pelmet. These can be made of stiffened fabric in a curved or geometric shape, or decorative perforated lengths of fibreboard can be bought and simply cut to size. Whichever you choose, you'll need to put up a pelmet shelf first.

This follows exactly the same method as putting up a shelf with a metal bracket, but this time it does not have to bear a heavy load, so smaller right-angle brackets can be used. The brackets won't be visible, so they need not be decorative.

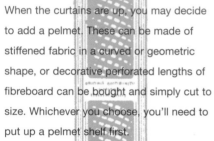

need

* 15-mm (⅝-in) thick MDF pelmet shelf cut to size
* 15-mm (⅝-in) wide hook-and-loop fastener (Velcro)
* staple gun
* screwdriver
* several right-angle metal brackets
* wall plugs and screws
* spirit level
* pencil
* drill with bit
* adhesive pelmet stiffener
* curtain fabric

STEP-BY-STEP PELMET

RIGHT-ANGLE BRACKETS   PELMET SHELF

CURTAIN RAIL

HOOK-AND-LOOP FASTENER

PELMET

TIP
The pelmet shelf needs to be about 10–15 cm (4–6 in) deep and a little wider than the curtain rail.

1 Cut a length of hook-and-loop (Velcro) fastener to fit around the side and front edge of the pelmet shelf. Separate the two halves of the hook-and-loop fastener, then staple one half to the edge of the shelf. Screw the metal angle brackets to the underside of the pelmet shelf.

2 Use a spirit level and a pencil to mark a guideline for positioning the shelf on the wall above the curtain track. Mark the screw positions, then drill the screw holes and insert wall plugs. Screw in the metal brackets so that the pelmet shelf is securely in position.

3 Cut a piece of pelmet stiffener to the depth you require and long enough to fit around the sides and front edge of the shelf. Trace your desired symmetrical curved or geometric shape along the bottom edge of the strip, leaving enough to fit around the sides at each end. Cut out along the design line.

4 Peel off the backing paper from one side of the pelmet stiffener to reveal the adhesive. Place the stiffener sticky side down on the wrong side of a strip of curtain fabric cut a little larger than the pelmet shape. Peel off the

backing paper from the remaining side, then make small snips into the excess fabric around the edge of the pelmet shape. Fold the resulting little tabs to the back of the pelmet and press firmly onto the adhesive. This will make a neat edge.

5 You can cut a piece of lining the same shape as the pelmet. Turn in the raw edges all round; then stick to the back of the pelmet. Glue or stitch the remaining piece of hook-and-loop fastener along the top straight edge of the pelmet, then attach it to the pelmet shelf.

And if you don't like curtains, you'll love these...

PUTTING UP 58 ROLLER BLINDS

Your decor may require a simpler look, or perhaps your windows are small. If that's the case, consider neat roller blinds or roman blinds instead of curtains. Both are available in kit form in various widths and designs. The kits come with full instructions and all the necessary fixings. Just choose the size closest to your window's dimensions; then trim to fit if necessary.

1 Measure the window recess carefully. Unroll the blind; then cut the wooden roller to size using a small hacksaw. Trim the blind material using a pair of scissors.

need
* hacksaw
* scissors
* bradawl
* screwdriver

2 Reconstruct the blind as per instructions. The material is usually fixed to the roller with a strip of double-sided adhesive tape. Simply peel off the backing paper to expose the adhesive, then press the top edge of the blind to it. Place the brackets in position on each side of the window recess, and mark the screw holes using a bradawl.

3 Secure each bracket in position, using the screws provided with the kit. These brackets are screwed directly to the wooden window frame, so there is no need for wall plugs. If you are attaching the blind to a masonry or hollow wall, then use the correct wall plugs.

* MEN ONLY WANT ME FOR MY DO-IT-YOURSELF SKILLS!

4 Push the plastic pin cap to one end of the roller blind, and the roller mechanism as shown to the other. Mount the roller blind onto the brackets as directed in the instructions. When the blind is in place, raise and lower it to make sure that it is working correctly and that the material has been fitted to the roller squarely.

Making a bookshelf 59

THIS PROJECT makes great use of crosshalving joints – a clever way of joining lengths of wood at a right angle. The basic principle is, you cut a slot as wide as the thickness of your material halfway across the width, then do the same with the other piece. The two then slot together in a cross shape. While considering this project I had an understair storage shelf in mind – hence the sloping shape. However, the more I thought about it, the more I liked the shape. Not only is it a great storage item with lots of compartments for organizing, but it has wheels, so you can easily move it around the room any time you want to reposition it.

1 When you buy the MDF sheets, ask the DIY store to cut them to the exact lengths to save you time. You can paint both sides of each piece before construction if you prefer, because it may be slightly awkward to paint when the bookshelf is complete.

2 Following the constructional diagram on page 185, outline the slots on each piece. Use a T-square and a pencil to make sure that the lines are straight and at true right angles to the edges of the cut pieces.

CHISELLING SLOTS . . . MORE FUN THAN THE BEST GOSSIP!

3 Place each piece securely on your workbench and cut very carefully along the marked lines using a jigsaw (remember to wear a dust mask while doing this). Chisel out the waste to make neat rectangular slots.

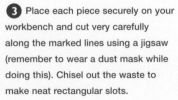

✳ GOT ANYTHING YOU WANT PAINTED PURPLE?

4 Following the constructional diagrams and template on page 185, begin to slot the pieces together. Glue end, side and base pieces together. If the slots are tight, then they will not require any gluing. Glue the end, side and base pieces to the slotted shelves.

need
* sheets of 15-mm (⅝-in) thick MDF cut into 20-cm (8-in) wide strips
* T-square and pencil
* jigsaw
* hammer and chisel
* wood glue
* drill with bit
* screw cups
* wood screws
* screwdriver
* satinwood paint in two shades
* mini paint roller and tray
* four castors plus screws

5 When the glue is dry, drill pilot holes and then screw the corners together. Slip a chrome screw cup over the screw before insertion. Using this decorative cover means that a countersunk hole is not required.

6 Using the mini foam roller, paint all the edges of the bookshelf and retouch any other areas. Screw a chrome castor securely to the underside at each corner of the unit so you can easily move it from place to place.

❋ ARNE JACOBSEN, EAT YOUR HEART OUT!

Fun with doors — take them off, put them on!

REMOVING A DOOR 60

Sometimes a door just gets in the way. If the door is never used, then there's no reason why it can't be removed – you could always replace it with a trendy bead or ribbon curtain! Then you can take the old door to the dump for some other person to use if they want to – recycling is a very good thing. A good-quality door can be expensive, and someone else might be very happy indeed to take your old one home.

need

* screwdriver
* shims
* wood filler
* filling knife
* sandpaper
* paint

Next time you pass a skip, look out for cast-off doors to die for.

❋ NOW WHAT DO I DO WITH IT?

❶ First, unscrew the hinges from the door and remove the door. You may need a friend to help you hold the door while you unscrew it. It is useful to place something firm under the door to reduce the strain on the hinges while you unscrew them, or to hold it at the correct height while you screw in hinges. You can use shims for this. They are small triangular wedges made of wood that you slip under the bottom edge of the door. You could also use magazines or thin wooden battens.

❷ Next, unscrew the lock striking plate on the other side of the door frame. Fill in all the screw holes and recesses with wood filler. When the filler is dry, sand the filled areas and repaint them to match the rest of the door frame. You may have to repaint the complete door frame if the finish looks a bit patchy.

HANGING A NEW DOOR

Changing a door can make a big difference in your home. Look out for second-hand doors that someone may have discarded – they may be nicer than the ones you have already. Remember to check the size before you commit. It is possible to cut a door to size but not to rescale a frame!

1 Remove the old door and use it as a guide for positioning the hinges on the new door. Place the doors side by side and mark the top and bottom of the hinge with pencil lines. You can make sure the guidelines are straight by redrawing them using a try square. Then measure the width of the new hinge and draw another guideline, again using the try square. Take your time at this stage to make sure that the hinges are marked accurately and correctly positioned.

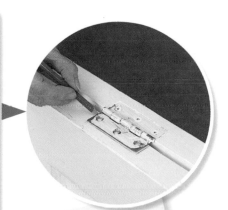

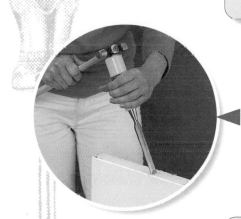

2 Use a hammer and chisel to make a shallow recess into which the new hinge will fit neatly. Tapping the blade of the chisel carefully, outline the recess first. Then make slightly deeper cuts in order to remove the waste.

The word 'unhinged' has acquired a whole new meaning!

need

* new door
* pencil
* try square
* new hinges and screws
* hammer and chisel
* drill with bit
* screwdriver
* shims

3 Place the hinge in the recess on the door and then indicate the positions of the screw holes. Drill pilot holes using your drill, then screw the hinge securely in place. Make sure that each screw is driven fully home and sits neatly in the countersink of the hinge.

4 Place the new door in the frame and support it with shims so that it is at the correct height while you screw it into place. You may be able to use the old hinge recess, in which case you can simply screw the new door in place. However, if the frame has been damaged or replaced, you may have to chisel out new recesses. Simply hold the door against the frame and mark the hinge positions; follow the same procedure as shown in step one to make the new recess.

More fun things to do with doors

CHANGING THE DIRECTION OF A DOOR

62

This may seem like a strange thing to want to do, but there may be a very good reason for doing so. You may have changed the purpose of a room, removed a wall or rearranged the furniture in a way that makes it more convenient for you to open the door on the opposite side.

need

* screwdriver
* chisel
* hammer
* surform planer

1 What you need to do first is to unscrew the hinges and remove the door from the frame. It may be possible just to flip the door over, chisel out both hinge and lock striking recesses on the opposite side of the frame and then rehang the door. You may be able to reuse the old screws.

2 It is unlikely that the door will fit exactly, so use the surform planer for fine-tuning. Otherwise, you'll have to take off all the hinges and the door handle, chisel new recesses, then refit everything on the opposite side of the door. See 'Hanging a new door' on page 97 to find out how to do this.

✳ SHE MAY HAVE BRAINS AS WELL AS BRAWN, BUT I'VE GOT THE LEOPARDSKIN PANTS!

TIP

This may or may not be obvious, but when you fill the screw holes in the other side of the door, don't close the door. If the spindle and the handles are lying on the floor on the outside and you're on the inside, that could cause a problem – you may not be able to open the door!

CHANGING DOORKNOBS

63

Does searching out new doors, drilling, filling and chiselling seem too much like hard work to you? Do the next best thing to give your tired old door a facelift. A new coat of paint might suffice, but new doorknobs are even better.

Remember, it's a door — not a battering ram!

① Loosen the screws on the old door handle on both sides of the door. Remove the handles to expose the spindle, then remove and discard the spindle.

need

* pair of new doorknobs plus screws
* screwdriver
* quick-drying filler
* filling knife
* sandpaper
* matching paint and paintbrush
* bradawl

② Fill in all the screw holes on both sides of the door with a quick-drying filler; sand smooth when dry. Apply a coat of matching paint to the filled areas to blend in with the rest of the door. You may need to repaint the whole door if the newly painted part looks patchy.

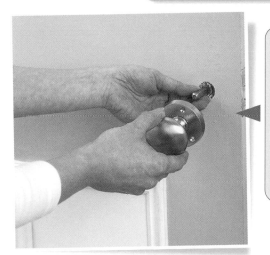

③ Insert the new spindle (doorknobs are sold in pairs, together with a spindle the right size). Slip the new dooorknobs onto the spindle. Mark the screw holes with a bradawl, then screw each new doorknob securely in place.

❋ NEVER MIND THE DOORKNOBS, NEXT TIME I'M CHANGING THE LOCKS!

Fantastic kitchen makeovers

KITCHEN WORKTOP MAKEOVER

Replacing a kitchen can be costly, but there are lots of inexpensive quick fixes that will improve the appearance of those tired old units. A simple paint job can do the trick for wooden cupboard doors and drawer fronts; sand down the surface, then repaint or stain in the colour of your choice. If the units are melamine, there are specially formulated products to use. Use only melamine primer and paint to achieve good results. When the primer is dry, apply one or two coats of melamine paint using a foam roller for a smooth finish.

THE INDUSTRIAL-LOOK KITCHEN LIKE THIS ISN'T DIFFICULT OR COSTLY TO ACHIEVE.

METAL WORK SURFACES COOL!

A copy of the industrial-look kitchen isn't too expensive to achieve. Most sheet metal suppliers will have cutting facilities and machines for bending neat corners. For a small fee they can cut, shape and bend a piece of metal to fit your worktop, and they can even drill the screw holes for you. Measure carefully and make a diagram, complete with all the details. All you have to do is screw it in place.

need
* stainless steel pieces to fit worktop or splashback or you could use aluminium
* drill with bit
* hammer
* centre punch
* wall plugs
* screwdriver
* round-head screws
* chrome screw cups
* small piece of timber
* clear sealing mastic

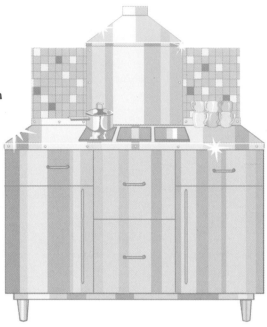

1 Carefully measure the kitchen surface that you want to cover; then make a simple diagram marked up with all the surface measurements so that the shop or sheet-metal supplier can get an idea of what is required. It's a good idea to ask the shop to drill some screw holes for you, but if you prefer to do it yourself, just mark the hole position and hammer through the metal with a centre punch. Place the metal on a small piece of wood before you hammer to protect the surface you are working on.

2 Place the worktop in position and mark the screw-hole positions on the wall and the edge of the work surface with a pencil. Drill the screw holes in the wall, insert the wall plugs and then make pilot holes in the edge of the work surface. Then screw the metal worktop securely in place. It is better to use domed screws and screw cups for this because they look more decorative. Seal all the edges with clear sealing mastic.

IDEAS

☞ If your worktop or splashback is tiled already, why not paint it a different colour? There are special tile-painting products available.

☞ Water-slide transfers are easy to use. Wet the transfer and slide it into position on the surface of the tile, then wipe away any excess moisture.

☞ You can use mosaic tiles to create a simple or complicated geometric design for your splashback.

❋ WEARING AN APRON LIKE THIS IS OPTIONAL!

STEP-BY-STEP PERSPEX SPLASHBACK — EASY

Tiles are an ever-popular choice for splashbacks, but if tiling seems like too much hard work to you, why not try a really quick alternative – perspex. It has the same properties as a tile in that it is waterproof and has an easy-to-clean surface. Perspex is also available in a wide range of colours to match the decor of your kitchen.

need

❋ coloured perspex sheet
❋ jigsaw (optional)
❋ drill with bit
❋ wall plugs
❋ screwdriver
❋ chrome mirror screws
❋ clear sealing mastic

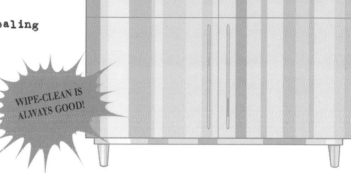

WIPE-CLEAN IS ALWAYS GOOD!

1 Measure the area to be covered; then either cut the perspex sheet yourself using a jigsaw, or ask the shop to do it for you. Most DIY stores have cutting facilities and will be happy to help you.

2 Make sure that the wall to which you will be attaching the perspex sheet is clean, dry and dust-free.

3 Drill holes through the perspex sheet at each corner at about 20-cm (8-in) intervals along each side. Hold the sheet up to the wall and mark the screw-hole positions on the wall. Drill the holes and insert the anchors. Then screw the sheet into position using mirror screws – these have decorative domed covers that screw onto the head of the screw.

4 To make sure that water does not seep down behind the panel, seal all the edges with clear sealing mastic.

TIP

Don't forget about handles! New handles can make all the difference in the appearance of your kitchen units. The choice is vast; these days all you have to do is take your pick. If you're lucky you may be able to use the same screw holes as those for the old handles.

Making a tongue-and-groove bath panel 65

TONGUE AND GROOVE derives its name from its shape. Essentially, tongue-and-groove planks have small tongue-shaped protrusions running along one side and a groove running along the other. The idea is that the tongue edge slots neatly into the groove of its neighbour, creating a panel with a decorative vertical pattern. Different profiles or cross-sections are available so that the finished panel can look more decorative. In the simplest form the grooves are just V-shaped, but there are more elaborate designs available.

1 I used straightforward V-profile tongue and groove for this new bath panel. This product usually comes in packs of six, but in various lengths and slightly differing widths. To calculate the amount you need, measure the length and height of the panel area, then check the dimensions on the packet. Divide the total length of the panel area by the width of a plank to get the number of pieces required, then multiply this figure by the height of the panel. Check the packets to see which one is closest to the amount you need – you can always buy a few extra planks.

❋ THAT WAS EASY. NOW I CAN HAVE A LONG SOAK IN MY COOL NEW BATH!

need

* tongue-and-groove planks
* pencil
* try square
* tenon saw
  sandpaper
* hammer
* panel pins
* satinwood paint
* paintbrush

2 First, prise off or unscrew all the old panelling and discard it. You will find that the bath is supported by a wooden frame, which is very handy for fixing the panelling. Measure the height of the bath panel carefully, then mark each piece with a pencil and try square to make sure the cut will be truly at 90 degrees to the edge. Check that the height from the top of the bath to the floor is consistent. You may need to cut each piece to a different length if your floor is uneven.

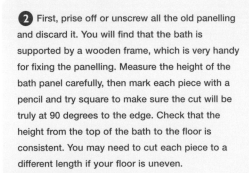

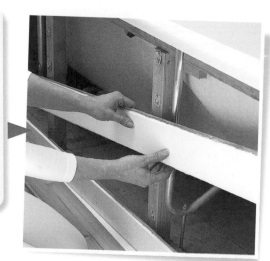

3 Place the panelling on a workbench or another sturdy surface. Carefully cut each piece to size using a tenon saw. Try to make sure that the cut is straight and square each time. Smooth off any rough edges at each end with a piece of sandpaper.

Tongue-and-groove planks

4 Beginning at the left-hand side, position the first piece with the groove edge flush to the wall. If there is a skirting board, you may have to cut the corner of the piece to fit around it. Hammer a panel pin into the tongue part at the top and bottom of the plank, securing it to the wood frame underneath. The idea is that when the next piece is in place it will cover the pin head. Continue adding one plank at a time in this way until the panel is complete. You will probably need to trim the final piece to fit.

5 Apply two coats of satinwood paint in the colour of your choice, allowing the first coat to dry thoroughly before applying the next. Paint in the direction of the wood grain, making sure to stipple the paint well into the grooves.

6 Jump into the bath and admire your handiwork. Your quick-fix bathroom makeover is complete!

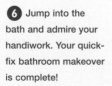

❋ QUACK! QUACK! CAN I GET IN NOW?

This will have you climbing the walls

This is a challenge, but the results are well worth it.

THE RULES OF TILING

Ceramic wall tiles have always been a popular choice for bathrooms and kitchens because they are durable, waterproof and easy to clean. However, there are no shortcuts in tiling – success depends on preparation and patience.

In essence, tiling involves spreading adhesive on the wall, sticking the tiles to it using plastic spacers so that the gaps are all even and then filling in the gaps with grout. For the first attempt, start with a small area, such as a kitchen splashback or a work surface. If this goes well, try a bigger project, such as the bathroom wall or shower cubicle.

PREPARATION OF SURFACES

Most surfaces can be tiled over, and if there are no existing tiles to remove, all you need to do is make sure that your surface is dry, clean, dust-free and flat.

Estimating the quantities

Most tiles will have an indication on the box of the approximate area the contents will cover. Calculate the area to be tiled, then divide by the figure on the box to get an idea of how many you will need to complete the job. Remember to allow for extra tiles for accidents and breakages!

REMOVING TILES

This can be a little more time-consuming. You will need a cold chisel and a heavy hammer to do this. If the tiles are old, they may be quite loose anyway, so you just need to prise them away from the wall. Otherwise you must crack each one with the chisel and then hammer and remove them piece by piece. It is important to remember that this is a job that requires safety goggles because the tiles will splinter and send small shards flying in all directions. To finish, rub down the wall and remove any old adhesive or grout to make a smooth surface for retiling.

LAYOUT

In general, it is better if the tiled area has a symmetrical arrangement of tiles, so that you can work from the centre outwards. This may not be appropriate in every case, however, because it's difficult to cut narrow pieces of tile to fit small gaps at the edges. Measure the area carefully. If the area is quite small, you could make a paper template and maybe shuffle the tiles around on it a little to find the best arrangement. If it's a larger space, make a tiling gauge. Lay out four or five tiles, complete with spacers, next to a wooden batten; then mark the positions on the batten. Hold the gauge up to the wall, and you will be able to estimate how many tiles will fit into a certain area.

TILING A SPLASHBACK

66

The wall area behind the taps in a kitchen, bathroom or utility room needs to be protected from splashes of water that could eventually cause damage to paintwork or wallpaper. Ceramic tiles are perfect for this because they are waterproof and easily wiped clean.

❶ When the surface preparation is complete, spread a layer of tile grout adhesive onto the surface to be tiled using a notched spreader. Press the spreader onto the grout, then drag it along to form little ridges. Don't put the adhesive on too thickly or the tiles won't stick properly. Work on a small area at a time, otherwise the adhesive will begin to dry before you reach it.

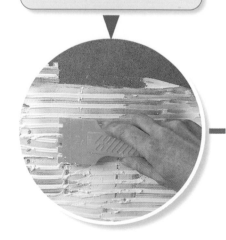

need

* tile grout adhesive
* notched spreader
* tiles
* tile spacers
* tile cutter
* paper
* pencil
* tile nibbler
* tile saw
* safety goggles
* flexible filling knife
* sponge

2 Press the first tile into position, then place the tile spacers at the corners. The spacers are cross-shaped, made of plastic and ensure a uniform space between the tiles. Continue laying all the whole tiles in this way, spacing each one carefully. These tiles are known as the 'field' tiles.

TIP

Buy plenty of extra tiles. Getting the hang of tile cutting requires a certain amount of practice! I know this to be true.

3 When you reach the edges of the tiled area, you'll need to cut to fit. Use a tile cutter for this. Measure the tile carefully, then place it on the base. Draw the blade along once, then use the integral tile cutter to snap the tile along the scored line. This is not necessarily as easy as it looks and does take a little practice – but if you persevere, you'll soon be an expert.

4 Make paper templates of any awkward areas such as around pipes, then trace around them in pencil on the tile. To fit around a pipe completely, cut the tile first, then draw the shape on either side of the cut edge.

5 Use a tile nibbler to remove the small areas. Squeeze the handles of the nibbler firmly and it will nibble away at the edge of the tile, enabling you to cut small shapes and curves in the tile.

❋ THERE, NOT SO HARD, WAS IT? STILL, THERE'S A BIT MORE TO DO ON THE NEXT PAGE . . .

6 Alternatively, use a tile saw to cut along traced lines. This tool has a round blade, so it's easy to follow curves. Place the tile on a workbench or table edge, then saw slowly along the traced lines. Take your time doing this; tiles are quite fragile and can break if handled roughly.

Filling knife

7 Position the tiles around the obstacle, which in this example is a pipe. First cut the tile either vertically or horizontally, then cut a semicircle into both pieces, which when fitted together forms a hole for the pipe.

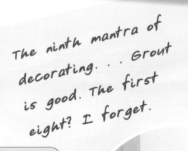

8 When all the tiles are in position and you are completely satisfied, apply grout to the spaces between the tiles using a flexible filling knife. Press the grout firmly into each space so there are no gaps. The plastic spacers are designed to be left in, so you need not remove them before applying the grout.

The ninth mantra of decorating. . . Grout is good. The first eight? I forget.

WARNING

☞ It is a good idea to wear protective eyewear when cutting tiles, just in case of stray splinters. Also, cut edges can be sharp, so take care not to cut your fingers.

9 Wipe away the excess grout with a damp sponge, then rinse the sponge and clean the tiles carefully. It is best to do this as soon as you have finished the job because grout can be difficult to remove from the surface of the tiles if left to dry out completely.

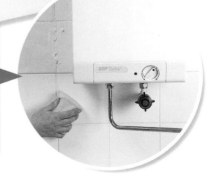

Now anything's possible!

TILING A SHOWER CUBICLE 67

The tiling principle is the same for small or large areas. When you feel confident, why not try a more ambitious project? When your shower cubicle is complete, you will probably want to add a few fixtures and fittings, such as a soap and shampoo shelf.

EVEN NORMAN BATES COULDN'T SPOIL THIS SHOWER!

Shower cubicle

WHOLE TILES ON THE OUTSIDE

HALF TILES ON THE INTERNAL CORNERS

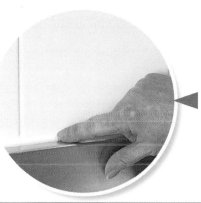

1 It is important to seal around the edges of any tiled area to prevent any water seeping underneath or behind the tiles. Silicone rubber sealant is specially formulated for this purpose. It is flexible and waterproof and available in handy tubes or guns. Simply squeeze into place, then smooth off neatly with a wet finger or the back of a spoon.

2 Each fixture will come with all the necessary fittings, such as screws and wall plugs. All you have to do is drill the holes. Ceramic tiles have a hard, slippery glazed surface, and you need to apply a piece of masking tape over the hole position so that the drill bit does not skid.

3 When the masking tape is in place, mark the hole position with a pencil cross and begin drilling. The masking tape will prevent slippage and allow you to drill neat holes.

ORDER OF WORK

1 The idea is to lay as many whole tiles as possible, then use cut tiles to fill in any gaps. If the cubicle has two sides, begin at the outside edges and work towards the internal corner. It looks better to have the cut tiles in the corner rather than along the outside edge.

2 If the area to be tiled has two corners, use the same method for the two sides, working from a whole tile at the edges towards the internal corners. For the back wall, establish a centre line and then work away from it in both directions towards each corner.

THIS IS A BIT OF A CHALLENGE BUT WORTH THE EFFORT!

Mosaic mirror splashback 68

IN MY OPINION, if a thing is worth doing, then it's worth overdoing. Why have a plain old mirror on your bathroom wall when you can have a wonderfully glamorous one like this? Mosaic tiles are available in a myriad of lovely colours and are relatively inexpensive to buy – and you really don't need that many. The ones I used are made of glass and are called vitreous tesserae. I also used mirror tiles, the type used in the bathroom. These had to be cut into strips and then into rectangles to fit the design. I hope this project inspires you to try a little mosaic-ing yourself!

need

* 61-mm x 91-mm (2½-in x 3½-in) sheet of 9-mm (⅜-in) thick MDF
* paper and pencil
* dust mask
* jigsaw
* oval bevelled mirror 35 cm x 45 cm (14 in x 18 in)
* double-sided heavy-duty mirror fixing pads
* tile nipper and cutter
* about 2 kilos (4½ lbs) mosaic tesserae in mixed blues
* six 15-cm square (6-in square) mirror tiles
* glass-cutting tool
* metal ruler
* tile adhesive
* grout and filling knife
* sponge and old toothbrush
* sandpaper
* screwdriver
* plate hangers plus screws
* drill with bit
* wall plugs

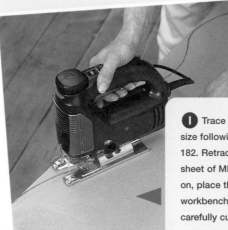

1 Trace the mirror template to size following the diagram on page 182. Retrace the outline onto the sheet of MDF. Put your dust mask on, place the MDF sheet on your workbench and use a jigsaw to carefully cut out the shape.

Little nippers

2 Lightly trace the position of the mirror on the MDF sheet and add basic design lines or choose your own random swirly shapes. Fix the mirror in place, using the double-sided sticky pads. Using the nipper, cut some tesserae into quarters. To do this, hold the tile in one hand, then 'nip' the tile at the edge to break it in half; then cut the halves in half. Cut more as you go along to fit the spaces in the design. For the mirror pieces, draw the glass cutter along the tile in a straight line (use a metal ruler), then snap into strips. Cut the strips into smaller pieces.

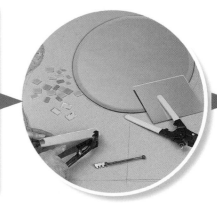

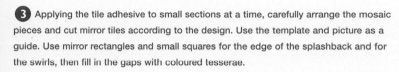

3 Applying the tile adhesive to small sections at a time, carefully arrange the mosaic pieces and cut mirror tiles according to the design. Use the template and picture as a guide. Use mirror rectangles and small squares for the edge of the splashback and for the swirls, then fill in the gaps with coloured tesserae.

4 Position all the tile pieces to your satisfaction, then set it aside and allow it to dry overnight. Apply grout to the mosaic with a flexible spreader to fill in all the gaps between the tiles, and also along the outside edge of the shape.

5 With a damp sponge, remove all excess grout; allow to dry again. Regrout any areas that you may have missed first time, then sponge clean. You will have to work quite hard now with that old toothbrush to clean any bits and pieces of grout from all the mirrored pieces so that everything is clean and sparkling. Finally, sand smooth the outside edge of the mirror.

6 Screw plate hangers to the back of the sheet; then mark screw-hole positions on the bathroom wall. Drill the screw holes, insert wall plugs and screw your new mosaic mirror splashback securely in place.

✳ APPLY LIPSTICK AND GO – THAT WASN'T SO BAD!

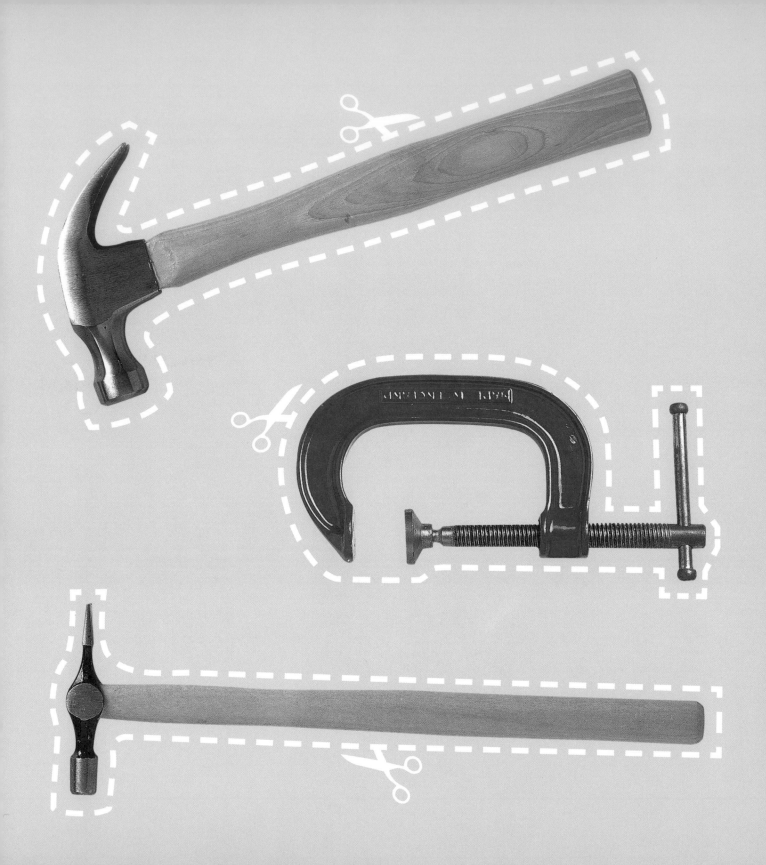

STORING IT

Keeping everything in its place! Ingenious ways to create storage space and brilliant projects to help keep your things neatly stowed away.

Put it away

GOOD ORGANIZATION is the key to success! I am one of those people who like things tidy – an 'a place for everything and everything in its place' sort of person. There's nothing more irritating, especially when you're in a hurry, than wasting time having to search through a pile of stuff for that important piece of paper, drill bit, shoe, spanner, pair of sunglasses, floppy disk, etc. In this section you will find lots of neat ideas for an even neater living space.

STORAGE `TIDY IT!`

You can put shelves just about everywhere – then by putting doors on them you can hide all the clutter inside. It's also easy to utilize wasted space under your bed or stairs and to make use of storage boxes – you can even make one that you can sit on (*see pages 120–21*). Even better is a storage box on wheels, so no matter how heavy it is, you can still push it from place to place.

ADDING SHELVES TO A WARDROBE OR CUPBOARD *69*

If, like me, you can't resist a bargain, you may have picked up an old wardrobe along the way. These tend to have hooks at the top rather than hanging rails, so to best use the space, add some shelves instead and use it as a clothes cupboard (or add hanging rails – *see opposite*). Try adding shelves all the way up the wardrobe, or add a single one at the bottom to stack your shoes more tidily.

❋ NOW THAT I'VE GOT MORE STORAGE SPACE, I CAN BUY MORE CLOTHES!

need

❋ MDF or wooden shelves cut to size
❋ pencil
❋ measuring tape
❋ softwood battens
❋ saw
❋ drill with bit and countersink bit
❋ bradawl
❋ screws
❋ screwdriver

❶ Measure the position of each shelf from the top of the cupboard, then mark with a pencil. Do this on both sides of the cupboard. Cut the softwood battens to size; then trim one end of each piece to a 45-degree angle. Mark with a pencil. You do not need to use a spirit level for this because it is not necessary for the shelf to be level with the floor, just square to the cupboard.

2 Drill clearance holes and then countersink holes about one third and two thirds of the way down the length of the wooden batten. Then hold the batten up to the pencil line. Now mark the screw-hole positions with a bradawl.

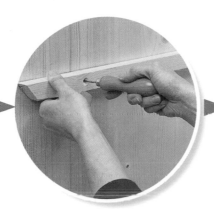

3 Screw each batten securely into position. When all the battens are screwed firmly in place, insert the shelves to complete the job.

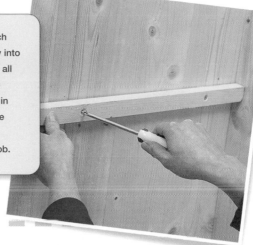

ADDING HANGING RAILS

The space underneath a shelf in an alcove, especially in a bedroom, can be a very useful space for hanging shorter garments such as jackets. Check out your local DIY store for chrome rails and end sockets. The rails are available in various lengths and can easily be cut to size with a hacksaw.

1 Place one of the chrome end sockets on one wall at the desired height. Mark the position of the screw holes with pencil dots. Repeat with the other socket on the opposite side of the alcove.

TIP
Use the same type of rail and fixing to add hanging rails to a cupboard.

2 Measure across the alcove, then cut the chrome rail to size with a small hacksaw, so that it fits snugly between the walls. Try the rail for size and make any necessary adjustments. At this point you can use a spirit level to check that the positioning is correct and that the rail is perfectly horizontal. Now drill each screw hole and insert wall plugs.

need

* chrome rail and end sockets
* measuring tape
* bradawl and pencil
* hacksaw
* drill
* wall plugs and screws
* screwdriver

3 Slip both of the end sockets onto the rail, and ease the rail into position. Screw each end socket into position securely to support the rail.

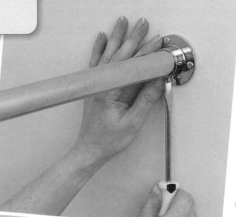

Making doors for open shelves

71

OPEN SHELVES, particularly in smaller living areas, can look cluttered and unsightly when heaped with piles of papers, old books or children's toys. The solution: hide them behind doors! Adding doors to shelves is a cheap and easy option, and an ideal vehicle for a little creative flair if you feel inclined. I had originally planned a 'yin and yang'-type design, but I got a little carried away. Now it looks more like 'image of woman' – well, in an abstract kind of way. You could always just make plain doors – the choice is yours.

need

* bookshelf
* sheet of 15-mm (⅝-in) MDF
* pencil and ruler
* jigsaw
* dust mask
* sandpaper
* drill with 2.5-cm (1-in) flat bit
* satinwood paint in two colours
* paintbrush
* four flush hinges
* bradawl
* screws

❋ NOW DON'T GET ALL EXCITED CUTTING CURVY SHAPES…A JIGSAW IS A POTENTIALLY HAZARDOUS POWER TOOL!

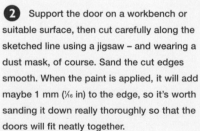

1 When you buy the MDF, ask the shop or timberyard to cut it to fit the front of the bookcase; this will save you a little time and effort. To begin, divide the MDF rectangle in half using a pencil and ruler. Next, sketch a pleasing wavy line approximately at the centre of the MDF. The two halves need not be symmetrical.

2 Support the door on a workbench or suitable surface, then cut carefully along the sketched line using a jigsaw – and wearing a dust mask, of course. Sand the cut edges smooth. When the paint is applied, it will add maybe 1 mm (¹⁄₁₆ in) to the edge, so it's worth sanding it down really thoroughly so that the doors will fit neatly together.

BREATHE DEEPLY AND APPLY STEADY PRESSURE… YOU'LL BE FINE.

Flat bit

3 Using the drill and flat bit, drill a finger hole in each of the doors. The positioning of these will depend on the shape of your doors. Roll a small piece of sandpaper into a tube shape, then sand the edges of the finger hole smooth.

4 Paint each door a different colour. Apply two coats, allowing each coat to dry before applying the next. Mark the hinge positions, and use a bradawl to make pilot screw holes. Screw the hinges to the door part first, then fix the doors to the shelf. Check that the doors open and close smoothly. If any part of the curved edge rubs or sticks, just sand down the problem area and repaint.

✳ PICASSO, EAT YOUR HEART OUT!

Radiators are shelving opportunities too!

BOXING A RADIATOR TO CREATE SHELVING SPACE 72

Radiators can be plain-looking objects – functional for sure, but not necessarily objects of desire. Radiator cover kits are available in various sizes and styles from mail-order companies. They arrive in flatpack form with full instructions and fixings. Most have perforated MDF panels in a wood or MDF frame. All the surfaces are unfinished, so you can paint the cover any colour you choose. Not only is the radiator now disguised, but you also have a surface for storage or display.

1 After studying the instructions carefully, you will have familiarized yourself with all the components and fixings. Check that you have the parts that you are supposed to have; then lay them all out on the floor and begin construction, following the instructions to the letter. Don't try getting clever with any shortcuts – you're bound to get into trouble!

need
* radiator cover kit
* rubber hammer
* screwdriver (kit will advise of additional equipment required)
* paint and paintbrush

* I DON'T THINK YOU CAN HAVE TOO MUCH PINK!

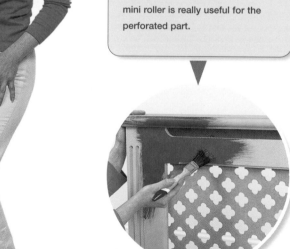

2 All the pieces are unfinished, so you can go ahead and paint the radiator cover your favourite colour. Use a paintbrush for flat parts or awkward edges, but a mini roller is really useful for the perforated part.

ADDING SHELVES TO RADIATORS

Radiator shelves are intended to prevent dark sooty marks from appearing on the wall behind and above the radiator. They also create a handy storage space or display area, especially in hallways, for keys or decorative objects. These shelves are available in different sizes and are sold complete with bracket screws and wall plugs. The metal shelf brackets are unusual in that they have posts that fit into drilled holes in the shelf. This way you don't have to remove the radiator in order to fit angle brackets.

need

* radiator shelf with fixings
* pencil
* spirit level
* drill with bit
* screwdriver

TIP
These shelves can be used in other areas too, because the brackets are concealed.

1 Draw a faint pencil line on the wall about 10 cm (4 in) above the top of the radiator to mark the position of the shelf bracket. Use a spirit level to ensure that the line is perfectly horizontal. Hold the metal bracket against the wall, matching it up carefully with the guideline. Mark each screw-hole position with a pencil. Remove the bracket. Drill the holes and insert the wall plugs. Use the screws provided with the kit to fix the shelf securely to the wall.

2 The three-pronged bracket will fit neatly into three holes drilled into the back of the shelf. Hold the shelf in front of the bracket, line up the prongs with the holes, then push the shelf into position. It will be held quite firmly, so there is no need for any additional screws or fixings. It is important to note that although the shelf is quite secure, it isn't a good idea to pile lots of very heavy objects on it.

3 Radiator shelves are made of MDF and usually have a melamine or wood-effect finish. To paint it, rub the shelf down with sandpaper, then prime and paint with products suitable for melamine.

✳ AT LAST, SOMEWHERE TO PUT MY GOLF TROPHIES!

✳ HMMM, YOU KEEP YOUR TROPHIES TO YOURSELF!

Now let your imagination run wild...

ADDITIONAL SHELVING IDEAS

74

The previous pages have described just a few basic methods of creating shelf space in your home. Yet there is even greater scope for your imagination when it comes to shelving – you are not just restricted to the material that you can buy from the DIY store, you know! Do a little hunting around in your local area for bits and pieces that may be interesting and useful. Look in skips for any discarded shelving material that you can revamp. Builders' yards are always a reliable source of materials to fuel your creative drive.

USING BRICKS AS SHELF SUPPORTS

Shelving does not have to be fixed to the wall or floor; it just needs to be solidly constructed. This is the perfect solution if you need some quick temporary storage space. Buy some basic pine shelves or cut some from MDF, paint them, then use ordinary bricks as supports at each end – and in the middle if the shelves are long. Stack the bricks in groups of four or more, depending on the size of the shelf required. Large grey breeze blocks also work well. Stack the shelves carefully to make sure they are balanced – and don't build them too high!

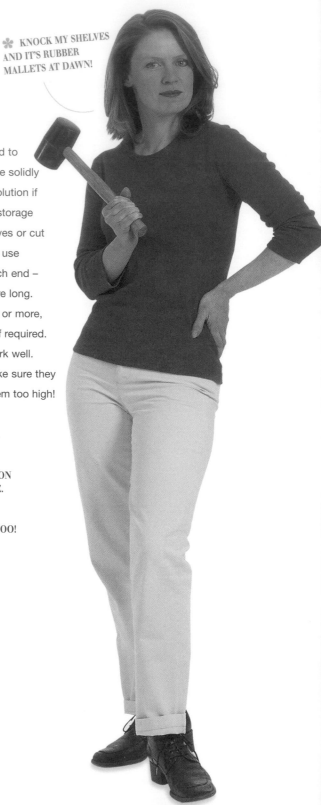

✳ KNOCK MY SHELVES AND IT'S RUBBER MALLETS AT DAWN!

A BOX FOUND IN A SKIP BECOMES A LINEN CUPBOARD WITH THE ADDITION OF A SHELF INSIDE. AND THE TOP PROVIDES EXTRA SHELVING SPACE TOO!

USING GLASS BRICKS

These transparent bricks have become very popular recently. They are frequently featured in style magazines in all sorts of interior environments. First available in clear glass, these bricks are now available in a range of colours. Use them individually as shelf supports.

SHELVES FROM RECLAIMED WOOD

If you've decided on the wooden batten and alcove approach to shelf creation, before you head for the MDF section of the DIY store, think again. The actual shelf part can be made from anything as long as it is straight and sturdy enough to take a bit of weight.

Consider old scaffolding planks, for example. Look in the telephone directory for a local scaffolding firm; they will usually sell you well-worn planks for relatively little money. Just cut to fit, sand off any rough edges, then balance across the battens in your alcove. You can leave the wood in its natural state, or stain or paint it.

Another option would be to ask a local timber merchant to cut you some shelves from seasoned wood that may be in the yard – if you're lucky, they will leave the bark on one edge. The bark will eventually fall off, but the edge underneath will have a really attractive natural shape.

Other little gems you may discover are old floorboards. Well-worn ones are full of character. Just rub them down and add a natural wax finish. Perfect.

Glass brick

COMBINE GLASS SHELVING WITH CHROME BRACKETS FOR AN ULTRA-CHIC LOOK. THINK ABOUT ETCHED OR COLOURED GLASS TOO.

A GIRL CAN'T HAVE TOO MANY SHELVES.

OTHER FABULOUS SHELVING IDEAS

☞ Glass shelves: these can be bought in kit form from DIY stores, or ask your local glazier to cut some to size for you. Tell the glazier that you plan to use the pieces for shelving, and he will supply you with toughened safety glass and polish the edges.

☞ Flatpack shelves: just customize to disguise their humble beginnings. How about metallic-effect paint, MDF borders or a paint finish?

☞ Large terracotta flowerpots: in a similar fashion to bricks, why not use these as shelf supports? They look great in the garden or on a roof terrace or balcony as shelving for plants. Leave them as they are (they will eventually patinate nicely if outdoors) or paint them with emulsion paint.

Making a storage-box seat

75

A STORAGE-BOX seat is a dual-purpose item – cram it full of stuff, then sit on it. The carcass is constructed from chipboard, a foam square is stuck on the top and then the whole thing is covered with some very smart-looking velvet. Add gold bun feet and a tassel just for fun. Check your DIY store for cheap offcuts of chipboard and ask them to cut them to size for you. See the exact dimensions in the box below and refer to the template on page 183.

EVEN EASIER IF THE SHOP CUTS THE CHIPBOARD TO SIZE!

1 Glue the four side pieces together using instant grip adhesive, to make a box shape 50 cm (20 in) square. Drill and countersink clearance holes near to the top and bottom and at the middle of each join. Drive in screws to secure the joins. Use a damp cloth to wipe away any excess glue that may seep from the joins after the screws have been driven in.

need

* 15-mm (⅝-in) thick chipboard pieces
* instant grip adhesive
* drill with bit and countersink bit
* chipboard screws
* wood glue
* 2 m (2 yd) velvet fabric
* needle and thread or sewing machine
* staple gun
* four wooden bun feet
* gold spray paint
* 50 cm x 50 cm (20 in x 20 in) piece of 10-cm (4-in) thick high-density upholstery foam
* 1 m (1 yd) contrasting colour lining fabric
* two brass flush hinges
* gold tassel

2 Glue the base to the sides to form the basic carcass. Drill clearance holes at 10-cm (4-in) intervals around the edge, and drive screws into each hole to secure the base. There is no need to countersink as these holes are on the underside of the box.

Dimensions

Base and lid
50 cm x 50 cm (20 in x 20 in) square

Front and back
50 cm x 30 cm (20 in x 12 in)

Sides x 2
47 cm x 30 cm (18½ in x 12 in)

Foam
50 cm x 50 cm (20 in x 20 in) square, 10 cm (4 in) deep

3 Cut a strip of velvet long enough to wrap around the box (including a 15-mm/⅝-in seam allowance), allowing about 10 cm (4 in) extra at the top edge and about 5 cm (2 in) at the lower edge. You may need to join fabric pieces to make the correct length. Join the pieces with right sides together to form a cylinder of fabric. Use a sewing machine or a small hand stitch to make the seams. Slip the velvet cover around the box – it should be a snug fit. Turn under the raw edge around the top edge; then wrap the edge around the top of the box. Use a staple gun to fix the fabric in place. Turn the box upside down and secure the bottom edge to the underside in the same way.

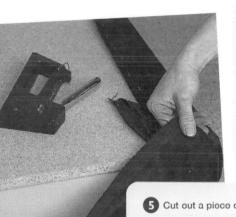

4 Spray the bun feet with metallic gold paint and allow to dry. In each corner of the box, drill a hole large enough to fit the threaded shaft. When you buy bun feet, you will also get metal washer-like fixings with gripping teeth on one side. Slip one over the shaft on the inside and, as you screw the foot in place, the washer digs into the chipboard and holds the foot securely.

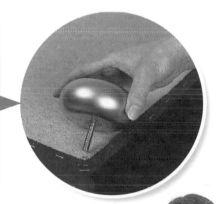

❋ WHO'S A CLEVER GIRL THEN?

5 Cut out a piece of velvet about 18 cm (7 in) larger all round than the lid. Place it wrong side up on your work surface, lay the foam on it, then lay the lid piece on top of the foam. Pull the fabric tight around each side of the lid, then secure using the staple gun. Pleat the fabric at the corners so that it forms a soft shape; trim away excess fabric at the corners to reduce bulk. Cut out a square of lining fabric a little larger than the lid, turn in the raw edge all around and staple or glue to the underside of the lid to cover all the raw edges.

6 Cut a strip of fabric to fit the inside and a square to fit the base with a 15-mm (⅝-in) seam allowance on all edges. Stitch the strip together to form a cylinder, then stitch to the base to form a square bag shape. Slip the bag inside the box. Fold 15 mm (⅝ in) over to the wrong side around the top edge. Then staple or glue to the inside. Screw in two flush hinges to join the lid to the box.

7 Screw the little gold tassel to the underside of the lid at the front.

Creating storage opportunities out of thin air

CONVERTING WASTED SPACE

76

❋ I JUST DON'T HAVE ANYTHING TO WEAR!

Take a look around you. Can you see any wasted space anywhere in your home? Most homes have plenty of nooks and crannies that can easily be converted to neat little storage areas of all sizes. It doesn't have to be square – with a bit of patience you can construct small units to fit any shape or space. Simply build a framework from timber battens; then cut a door to fit and attach with hinges. When you've filled it with new clothes, you can start thinking about more storage space!

ABOVE DOORS

Raise your eyes skywards – any room for shelves above the doors or at picture rail level? This is a great idea if you have lots of books that you don't need every day but are reluctant to discard. It also makes a nice display area for all those bits and pieces you've collected over the years.

CORNERS

These are neglected areas in my opinion. You can construct your own L-shaped shelving system or corner cabinets, or choose from many that are commercially available. Scour second-hand shops for sturdy units to customize or cut to fit.

USING HOOKS AND RAILS

☛ Screw cup hooks to the underside of shelves and wall-hung cupboards to create extra space for coffee mugs, jugs or kitchen utensils.

☛ Add towel rails to the sides of kitchen units or under shelves for dishcloths or for hanging a dustpan and brush or other household items.

☛ Look in the DIY store for large heavy-duty hooks for hanging up folding chairs, a bicycle or a ladder when not in use.

UNDERBED STORAGE BOXES HIDE A MULTITUDE OF CLUTTER.

HANG THINGS ON THE WALLS WITH HEAVY-DUTY HOOKS AND RAILS.

UNDERSTAIR STORAGE

The triangular space under the staircase has traditionally been the home of brooms and buckets, vacuum cleaners, spare lightbulbs and old overcoats. That's fine if you have a door to hide it all. But think again – it's a waste of valuable space really, and I'm sure there's somewhere else for the vacuum cleaner and brooms to live.

MINI OFFICE

The number of people choosing to work from home these days is increasing rapidly, but there's often no space for a complete office. Maybe what is needed is a compact work station under the stairs. Then, if you really need some separation between your work and home environments, simply hang up a curtain or some shaped fabric panels, and when the work is done, the fun can start.

Mind the slope! Don't jump up from your chair in a hurry!

✳ MMM, THIS ONE IS DEFINITELY A WASTE OF SPACE!

TIP

If you don't have enough room for a mini office, your understair nook might be the ideal place for shelves for audio equipment or a television, or even that extensive vinyl record collection you just don't want to get rid of.

$ $

Underbed storage project 77

THAT SPACE underneath the bed is a vast storage opportunity – so do not waste it. Tucking everything you possibly can under the bed in a haphazard fashion makes things difficult when you need to go and retrieve something that has worked its way right to the back. What you need is something with wheels! All you need to do is construct a shallow underbed-sized box, or two to go side by side, then fit a lid (anti-dust device) and put it on wheels. You can wheel it out, lift the lid, retrieve the item and wheel it back again.

need

* 15-mm (⅝-in) thick and 6-mm (¼-in) thick MDF cut to size
* softwood battens
* saw
* instant grip adhesive
* drill with bit and countersink bit and 2.5-cm (1-in) flat bit
* screwdriver and screws
* wood filler and filling knife
* sandpaper
* hammer and panel pins
* four castors plus screws
* satinwood paint (two colours) and paintbrush

1 Cut the softwood battens into four pieces 6 mm (¼ in) shorter than the height of the box side pieces. Use instant grip adhesive to attach a batten at each end of both short side pieces. Position the battens flush with the bottom edge, leaving the 6-mm (¼-in) gap at the top; this forms supports for the box lid as well as reinforcing the frame of the box.

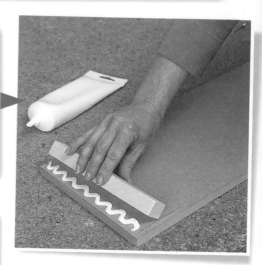

2 Drill and countersink clearance holes, and then drive in screws to attach the battens to the side pieces.

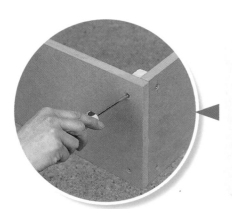

3 Glue the short side pieces to the long front and back pieces to form a rectangle. Drill, countersink and screw the pieces together as before.

4 Use a filling knife and wood filler to fill each of the countersunk holes. Sand smooth when the filler is dry.

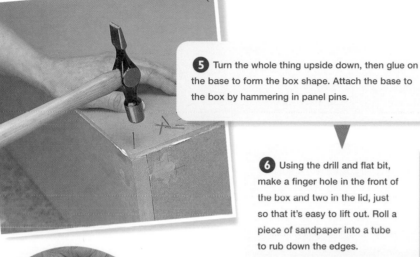

5 Turn the whole thing upside down, then glue on the base to form the box shape. Attach the base to the box by hammering in panel pins.

6 Using the drill and flat bit, make a finger hole in the front of the box and two in the lid, just so that it's easy to lift out. Roll a piece of sandpaper into a tube to rub down the edges.

7 Turn the storage box upside down and then, using the screws provided, fix a chrome castor to each corner of the underside.

✻ AND HERE'S ONE I MADE EARLIER!

8 Turn the box the right side up and apply two coats of paint to the outside and to one side of the lid, allowing each coat to dry before applying the second coat. Paint the inside of the box and lid a bright contrasting colour.

Dimensions

Cut from 6-mm (¼-in) MDF
Base x 1
61 cm x 90 cm (24 In x 36 In)
Lid x 1
58 cm x 87 cm (22¾ in x 34¼ in)
Cut from 15-mm (⅝-in) MDF
Long side x 2
90 cm x 20 cm (36 in x 8 in)
Short side x 2
58 cm x 20 cm (22¾ in x 8 in)

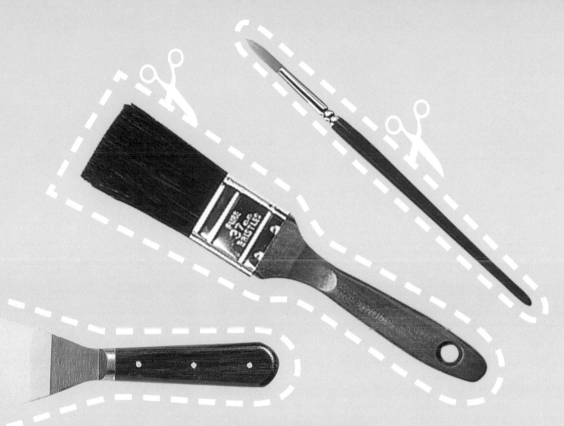

FAKING IT

Now for the fun part — decorating! Make your house a home with decorative tricks — paper it, paint it, rag it, bag it, stencil it or stamp it.

Painting and decorating

BY NOW YOU will probably have fixed all the broken floorboards, tended to the leaky pipes and replaced the old doors in your home. However, there's still plenty to do. There's quick-fixing, wallpapering, painting, paint effects and all of the preparation that this entails. Basic DIY is probably the most straightforward part of home maintenance, because when something is broken or doesn't work very well, you tend to know about it – and now you know how to fix it. Decorative projects involve decision-making in more of an emotional sense. The problem is not necessarily how to do it, but what to do (then how, of course).

CONSIDERING THE PROJECT HA!

The chances of having money to spare after you've just purchased a new home are slim. However, you shouldn't allow money, or lack of it, to stem your creative flow. Careful consideration is needed now: what is essential, what can you live with and what can you quick-fix and fake? A well-papered room can easily and inexpensively be given a quick paint facelift. This can free up the bank balance for a more important and costly decorating job. The secret is not to spend your budget all at once. Consider your home as a whole and do things step by step.

FOOLS RUSH IN...

Style is a wonderfully individual thing. Think of your home as a blank canvas and do anything you want to it. If your ideas are clear, then you can skip the 'what' and move swiftly to the 'how'. But if, like me, the idea of having your own place, no money and a free rein proves overwhelming, then perhaps you need to do some serious considering. In a fit of panic and poverty it is perfectly feasible to paint everything white, and then strategically place a large stone with a hole in it and a piece of driftwood on the window sill (as I did) in the spirit of modern minimalism. In reality, of course, I couldn't live with what I

DECIDING ON THE COLOUR SCHEME IS THE FUN PART! GET YOUR INSPIRATION FROM BOOKS AND MAGAZINES.

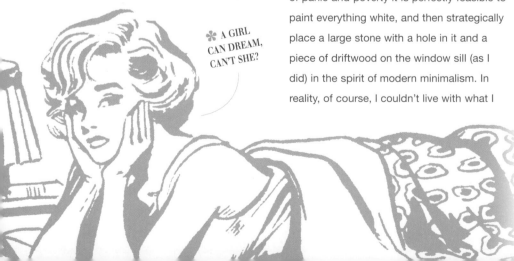

✻ A GIRL CAN DREAM, CAN'T SHE?

had done and couldn't decide what else to do or what colour to do it!

No matter how desperate you are to change everything all at once, you should take time to consider what it involves. Do you have the time to finish the job or not? For example, floral embossed wallpaper may look fine when painted a bright colour as a quick fix, but are you going to have time to finish what you started? Do you really want to live with half-stripped, nasty wallpaper with a patchy pink plaster combination for a month until you have the chance start work again and complete the project?

WHAT'S UNDERNEATH?

In some houses, particularly those that are a bit older, it is advisable to check the wallpaper before you begin to remove it because cracked and crumbling plaster might come off too! You may end up with a big replastering job on your hands, so don't start ripping and tearing like crazy.

Everyone would love to have perfectly smooth walls. Some are and some just aren't. In general, if the walls are slightly uneven, wallpaper or lining paper is the answer. But if the imperfections are minimal, you can get away with careful filling and then painting straight over the top.

The top layer of vinyl wallpaper can sometimes be carefully peeled off, revealing a perfect surface for painting. The decor in my bedroom really was not to my taste, so that's exactly what I did. I then painted it white (modern minimalism again). Brilliant quick fix. If you intend to try one of the paint effects

discussed on the next pages, don't worry if the wall surface is not perfectly smooth because a slightly uneven surface isn't going to show up that much.

PLANNING AHEAD

It is a good idea to sit yourself down with a piece of paper and make a plan for your decorating project. Alternatively, invite your friends round and have a brainstorming session. Buy piles of style magazines, get inspired and then make some sketches. Go to your local DIY store and get some lovely paint shade cards or tester pots. Paint patches of colour here and there on your walls to help you decide which ones you really like. It's well worth making an effort to get it right. This is your home, your sanctuary and your environment so you want it to look the best that it can.

PINK, PINK, PINK, PINK AND MORE PINK!

BUT CAN YOU LIVE WITH IT 24 HOURS A DAY, 7 DAYS A WEEK?

Grab your tools and let the preparations commence!

ORGANIZING YOURSELF

Time flies when you're engrossed in a DIY project. How long a job is likely to take should be a major consideration when you plan your work schedule. You can fit half-hour or hour-long jobs easily into an evening or a bit of spare time at the weekend. Even a big job can be split into small sessions, providing that it is in a place that you don't have to use every day. If there's a room with lots of laborious paint-stripping to do, for example, it's easy to do it a little bit at a time if you can just close the door on it. However, nothing will drive you crazier than having to step over lots of equipment in the kitchen or bathroom every day for weeks on end, when you have only an hour or two every weekend for working on the project. It's worth planning ahead to free up a few days at a time to finish a job completely – you'll also feel a great sense of achievement afterwards.

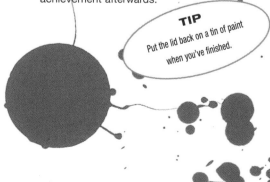

TIP
Put the lid back on a tin of paint when you've finished.

PREPARATION

Confucius certainly knew a thing or two about preparation: 'In all things, success depends upon previous preparation, without which there is sure to be failure.' It really does pay off – I can't stress it enough. Think the job through carefully beforehand, and get all your tools and materials organized so you don't have to rush to the DIY store mid-project for that crucial item, without which there is sure to be failure!

In general, you should start at the top and then work down. For example, paint the ceilings first, then the covings and woodwork and then begin the papering or painting. Flooring will probably be the last on the list. However, there are exceptions. If you intend to sand the floor, then do that first – it would be foolish to do all the decorating, then have everything covered with dust.

The same principles should be applied to tiling; you shouldn't tile over wallpaper, so install and grout the tiles before starting any

papering or painting. Also, try to anticipate any intrusive DIY projects. If a big hole in the wall is needed to accommodate a pipe or other fixture, it wouldn't be a good idea to decorate beforehand. Take care of the really dirty jobs first: don't lay beautiful carpet in the hallway, then drag a dusty floor-sanding machine over it to get to the living room. Choose a logical approach – then everything will fall neatly into place.

Remember that things always take longer than you expect. Allow for drying times for paint, wallpaper, filler and all the cleaning up afterwards. Try to organize a job that will take a few days to be carried out in a logical sequence. Aim to complete painting or hanging the lining paper at the end of the day so it can dry overnight; then the following morning you'll be ready for the next stage. Before painting, make a note of whether the paint is oil- or water-based. Oil-based paints tend to take a lot longer to dry, so you will need to take this into consideration if you will be applying several coats.

Time is also something that you should consider if you need to call in a professional – you pay for every last minute of their time (including tea breaks). If you can do the laborious preparation, room clearing and anything else that will take a few hours, do it. Then the professionals can come in and do a fabulous job unhindered.

Specialist brushes

After all that hard work, here are some fantastic make-up tricks!

SPECIAL EFFECTS

Antiquing

☞ Antiquing effects use an abrasive technique, either to rub away a top coat of paint to reveal the undercoat or to imitate wear and tear.

Colour washing

☞ This is a delicate paint effect that is created by applying a watery paint mix over a base colour using a sponge or brush.

Dragging

☞ Fine lines are created on a surface by dragging a flogger brush through a top coat of paint mixed with scumble glaze over a base coat.

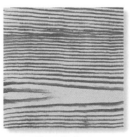

Graining

☞ A specially designed plastic graining tool is used to create a wood-grain effect. This is great for plain surfaces such as MDF or melamine.

Marbling

☞ Marble effects can be achieved by using artists' brushes to make the coloured veins, then a large brush to blend the background colour.

Ragging

☞ Apply a base coat, then a top coat of paint and scumble glaze. Simply work a rolled-up rag over the surface to make a random pattern.

Sponging

☞ Use a natural sponge to apply paint over the base coat. The irregular holes in the sponge will create a subtly mottled pattern.

Stippling

☞ A contrasting top coat is applied, using a large brush to create a pattern of tiny dots. Concentrate the paint in some areas for a mottled effect.

Now there's the rub — preparing to make wood look good

PREPARATION FOR PAINTING WOOD

78

Preparation is the key to success. As tempting as it may be to apply gloss paint over existing paintwork on doors, skirting boards or door frames as a quick decorating fix, the end result will almost certainly be inferior and ultimately disappointing. Paint applied without any preparation will peel off in no time and you will just have to do it again – in the end preparation saves time. Painted woodwork can also discolour over time, and if you plan to repaint or paper the walls, that grimy old paintwork is really going to show you up.

If the existing paint is fairly sound, it is perfectly feasible to rub it down carefully to take off the shine and create a good key for a fresh coat of paint. This can be a tough job, whichever way you look at it. Rubbing down or sanding by hand is hard work, but some areas just have to be done by hand.

SANDING BLOCKS

☞ When sanding flat areas or square edges of woodwork, it is best to use a sanding block. This is just a block of wood or cork around which you wrap a sheet of abrasive paper. Fold and tear the sheet in half, then wrap it around the block. If you need to sand moulded or curved areas, ease the paper into the shape using your fingers or a shaped block, or cut a piece of dowel and wrap the paper around that. Flexible sanding sponges are good for shaped surfaces too; just wet before use, sand the

surface, then rinse the sponge when it becomes clogged. Sandpaper is available in various grades ranging from fine to coarse. Start with a coarse grade, then work down to fine.

> **Abrasives**
> Sandpaper is available in various grades from fine to coarse; you can buy assorted packs to see you through most woodworking tasks. The finer the paper, the finer the finish.

✳ THERE'S NO NEED TO BE SO COARSE!

POWER SANDERS

☛ These are wonderful inventions. Detail sanders have a handy pointed sanding plate for use over small areas or in awkward corners. For bigger areas, however, it is advisable to invest in a larger machine such as a belt-driven or orbital sander. Another important thing to mention is dust. Sanding in any shape or form creates dust, so wear a dust mask. Some machines have an attachment that can be fitted to a vacuum cleaner. When using a sander, pass it over the work in regular forward and backward sweeps with the grain of the wood, overlapping each movement with the next. With a detail sander you will need to manoeuvre it to reach all the complicated areas, then attack any leftover bits by hand.

If you want to paint metal, the same rules apply. Rub it down well first with abrasive or wet and dry papers designed for metal, then paint with specially formulated paint.

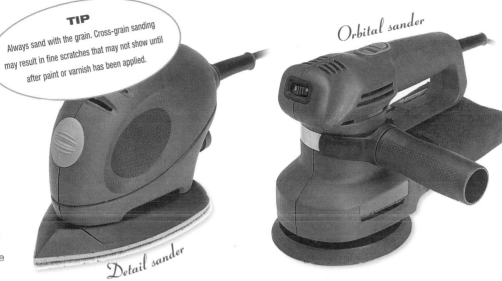

TIP
Always sand with the grain. Cross-grain sanding may result in fine scratches that may not show until after paint or varnish has been applied.

Orbital sander

Detail sander

GET IT OFF!

STRIPPING PAINTWORK

☛ You may be unfortunate enough to have multiple layers of uneven gloss paint to contend with, or lots of cracks and chips to repair. The answer in this case is to strip! Again, this isn't a quick job, but it's worth doing properly to achieve a good result. Alternatively, if the chips are minimal and the paint is relatively smooth, just fill, sand and then repaint. The two methods of stripping paint are as follows:

HOT AIR GUN

This power tool emits a fierce heat. Pull the trigger and point at the area to be stripped. The heat melts the paint, which can then be easily scraped off with a flexible knife.

IT'S JUST FULL OF HOT AIR!

CHEMICAL STRIPPERS

Chemical strippers are caustic and the fumes are unpleasant, so always wear protective gloves and work in a well-ventilated area. Brush the stripper onto the area as directed, and then leave for the required time. The paint will soften and bubble, by which time it will be easy to remove with a knife or shave hook. Use a small stiff brush or wire wool for delicate areas. Neutralize the surface with white spirit (or the substance recommended in the instructions) before repainting.

Decisions, decisions... wallpaper can be glamorous too!

WALLPAPER TYPES

THE BIG COVER-UP

Wallpaper is a general term to describe wallcoverings of all types. If the thought of wallpaper puts you off, think again. It's not all unpleasant embossed motifs and chintzy florals. Wallpaper manufacturers now offer a huge selection of beautiful designs and textures to choose from. Most have co-ordinated ranges with plain colours, tasteful patterns and borders designed to complement each other. The hard part is deciding which style to pick.

WHICH ROOM?

In most cases the colour, pattern or texture will be the driving force behind your decision, but there are other factors that you should consider too. Some wallpapers are specially designed for kitchens or bathrooms. Others are more delicate and should be used for decorative effects only in a bedroom or the living area, where they are not exposed to dampness or steam.

Papers for kitchens and bathrooms have a plastic wipe-clean coating to repel moisture created by cooking steam and to ease removal of dirt and grease. Always check the manufacturer's guidelines on the packaging before you buy wallpaper; most will suggest the best use for that particular paper.

TEXTURED WALLPAPER WILL DISGUISE UNEVEN PLASTERBOARD OR PLASTERWORK.

WALLPAPER COMES IN ALL KINDS OF PATTERNS AND STYLES. GO CRAZY WITH STRIPES!

EVEN CREATIVE TYPES NEED TO KNOW THEIR MATHS!

Estimating the quantities

The standard wallpaper roll is about 520 mm (1 ft 9 in) wide and about 10.05 m (33 ft) long. The basic calculation you need to make is this: measure the drop from ceiling to floor; divide the length of the roll by this measurement to calculate the number of drops per roll. Then measure right around the room; divide this measurement by the width of the roll to calculate how many drops you will need. Multiply the total number of drops by drops-per-roll figure to estimate the number of rolls you will need. The same principle applies to the ceilings: just look at it as though it is a big wall, the drop being the length from one end to the other. You can measure this on the floor.

❋ MMM, NOW, WOULD LIME GREEN WITH PURPLE SPOTS BE TOO MUCH?

If you ever wondered how granny got 'the look', here it is!

OKAY, IT'S USEFUL, BUT PLEASE DON'T!

Lining paper

☞ Inexpensive natural-coloured paper applied to walls before hanging heavy wallpaper or used as a base for painting if the walls are not perfectly smooth. Lining paper can cover a multitude of sins and is available in various grades.

Woodchip paper

☞ This type of wallcovering was popular in the 1970s. Woodchip wallpaper covers everything, including uneven walls, and is usually very difficult to remove. Not particularly attractive but handy and inexpensive. Must be painted.

Washable paper

☞ Plain or patterned flat paper that has a coating of white glue applied to the pattern side. This renders the surface waterproof and wipe-cleanable.

Vinyl paper

☞ Patterned or plain paper (or in some cases cotton) with a clear vinyl coating. Heavy-duty vinyl papers are very useful for kitchens or bathrooms. Blown vinyl paper has a relief pattern embossed into the backing, which is then coated.

Printed paper

☞ Flat paper with multicoloured designs printed on one side. There is a huge range to choose from at different prices. Set the budget first, then make your choice. Most patterns are machine-printed, but there are beautiful hand-printed wallpapers available – these are lovely, but expensive.

Ready-pasted paper

☞ This type of paper is ingenious. It comes with a pre-pasted backing, so all you have to do is soak each roll in the specially designed trough filled with water, lift it from the trough and hang it directly on the wall. This will save you a great deal of time and messy pasting.

Embossed paper

☞ Plain or patterned paper with an embossed pattern that can create stunning effects. Plain papers can be painted over. Sometimes viewed as the type of wall covering seen only in restaurants and hotels, the backing paper has fibres glued to the surface to form a raised pattern.

Borders, friezes

☞ Patterned, plain or embossed bands of paper designed to run horizontally at ceiling or dado rail height. Applied after the paper has been hung.

Getting down to the bare necessities

Easy if it's vinyl, but if it's woodchip, then it'll be tough going!

STRIPPING WALLPAPER

 79

So you've decided that the ugly old wallpaper is coming off and you are going to start from scratch. There are a few essential tools that will save you time and effort during the wallpaper-stripping process. The first is an orbital scorer – an ingenious triangular device with spiked wheels that makes small perforations in the wallpaper. The second tool is a wallpaper steamer. Both of these are inexpensive to buy or hire and they really cut down on the hard work.

It is well worth spending a little time ensuring that all the wallpaper to be stripped is well perforated before soaking or using a steamer. This ensures that the moisture penetrates the wallpaper thoroughly and dissolves the paste underneath, thereby facilitating easy removal. Use the orbital scorer for this – just run it in a random fashion all over the surface of the wallpaper. You can also score the paper with the edge

of the stripping knife in a cross-hatching pattern, but this can leave scratches in the wall. You will probably have plenty of holes and imperfections to fill in after the wallpaper has been removed, so you don't want to make more work for yourself by creating scratches you'll have to cover up too. So if you do decide to use this method, be careful.

❋ NO, WE DIDN'T MEAN THAT KIND OF STRIPPING!

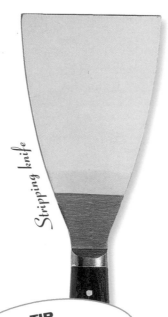

Stripping knife

STRIPPING WITH WATER

☛ Fill the bucket with warm water; then use a large brush or sponge to apply the water to the wall, thus soaking the wallpaper. As the paper absorbs the water, it becomes soft and easy to remove with the scraper. Work on a few square metres at a time or the paper will start to dry out again before you have a chance to remove it.

☛ Make sure that you remove all the tiny bits of paper while they are still damp. Don't wait until the end or they will have dried out and you will have to soak the wall again.

TIP
Try making your own scorer by hammering nails into a block of wood so that the ends of the nails stick out the other side.

need
* orbital scorer
* wallpaper stripper or water bucket and sponge or brush
* stripping knife
* bin bags

STRIPPING WITH A STEAMER

I strongly suggest that you try to get a steam wallpaper stripper – I guarantee that you will not be disappointed. This really is a fantastic time-saving device. It looks like a cross between a kettle and a vacuum cleaner. You need to be careful once the water has boiled because the steam is very hot and can scald your hands. Don't use the steamer for more than around 45 minutes, and never leave it switched on and unattended.

1 The tool pictured on the right is called an orbital scorer. It has three spiked swivel wheels on the underside, which pierce thousands of tiny holes through the surface of the wallpaper. Hold the handle firmly and work over the walls in random circular patterns. This is quite an aerobic task but it will be worth the effort because the paper will be very difficult to remove if it has not been properly scored beforehand.

2 Following the manufacturer's instructions carefully, fill the reservoir with water, then plug in the steamer. Wait for the required amount of time until the water boils. Place the steam pad on the wall and hold it in the same place for about 30 seconds. This should be enough time to make the wallpaper easy to remove. Thicker papers, such as woodchip paper, may require a little longer. Try a small area first to give you an idea of how long to leave the pad on the wall.

Keep the room well ventilated when steam-stripping to avoid the sauna effect!

3 Remove the soft bubbled paper, using the stripping knife or scraper. Throw the stripped paper into bin bags as you go along. Don't be tempted to leave it in soggy piles on the floor until the end of the job, because the paste will dry out again and the bits of paper will have stuck to everything – floor, dust sheet, skirting boards.

* I LIKE MY LUMPS AND BUMPS, BUT NOT ON MY WALLS!

Filling in the cracks and gaps for a super-smooth wall

PREPARING A WALL

Whether you intend to paint or paper your walls, it is necessary to prepare them first. That means filling in any holes or cracks, and smoothing out any lumps or bumps. It would be a rare occurrence, actually a miracle, if you were to strip your walls and find perfectly smooth plaster underneath. Even in homes that are only a few years old, you will still find holes from picture hooks or slight damage from moving furniture, for example.

Do not hurry through the filling process, any spots that you've missed or not sanded down properly will show up later, especially if you choose a paint finish. A good tip that I learned from a decorator was to roll a coat of thinned white emulsion paint onto the walls before filling. This shows up all the holes, cracks and bumps very nicely for you. It is known as a 'miscoat'.

Take a good look at your walls when they are stripped and bare. Is there any damage from leaks or areas where the plaster is loose? It's unlikely that your walls will be completely free of holes and cracks, but if they are in bad condition, this may mean a phone call to a professional plasterer. It may be smart to have the walls resurfaced. Plastering is a big job and is best left to the professionals. Small areas you could tackle yourself, of course!

YOU DON'T WANT TO BE WITHOUT THIS STUFF!

Filler

FILLERS

☞ For most imperfections a general, all-purpose decorator's filler or fine surface filler can be used. Both come ready-mixed in a tub or a tube. These are easy to use and quick-drying, but remember to put the lid on the tub when you finish or have a coffee break; it really does start to dry quickly.

For areas that are likely to experience movement, such as the architrave surrounding a doorway, use a flexible filler to prevent possible cracking.

1 First, brush off any debris or loose particles from the areas to be filled. If the hole is large, dampen the edges with a wet sponge. This lengthens the drying time of the filler, so it is less likely to shrink when it dries. The moisture also improves the bond between the filler and the plaster in order to prevent the filler shrinking after application.

2 Scoop a small amount of filler onto the filling knife and apply to the hole or crack. Press the filling knife firmly against the wall, spreading the filler fairly flat and pushing it into the hole. You may need to draw the blade of the knife over the imperfection a few times to make sure that it is completely filled. Fill all imperfections in this way and then allow to dry. If you have deep holes to fill in, twist a strip of newspaper tightly, then use the filling knife to force it into the crack; or screw a piece up into a ball to fill a hole. This provides a base so that you can apply filler over the top.

✳ I WONDER WHAT HE'S UP TO WHILE I'M HERE DECORATING?

TIP
For deep holes or cracks, bunch up a piece of newspaper and push it firmly into the hole or crack using the tip of the filling blade or your finger. Fill over the top as before.

3 When the decorator's filler is completely dry and hard, sand it smooth with sandpaper and a sanding block. You could use a power sander if you have a larger area to cover. Run your hand over the sanded area to make sure it is completely flush with the wall. If necessary, apply another layer of filler and repeat the process. If you need to apply a second layer, first wipe off any dust from the first sanding with a damp sponge or cloth. Sanding is a dusty process; if you decide to use a power sander, remember to wear a dust mask.

need
* sponge
* all-purpose filler
* filling knife
* sandpaper

You've got the old stuff off, now put the new on

WALLPAPERING 81

This is a big project, so make sure you have a free weekend to do it in!

need

* wallpaper or lining paper
* measuring tape
* plumb line or spirit level
* pencil
* wallpapering scissors
* wallpapering table
* string
* wallpaper paste
* paste brush
* bucket
* step ladder
* wallpapering brush
* seam roller
* craft knife

The stripping, sanding and filling is complete and the woodwork is painted. Your walls are now in a fully prepared state and ready for papering. Wallpapering is a job that can be completed quite successfully solo – but if your walls are expansive and the ceilings high, you may want to try asking some friends to help you out. Besides, you can always reciprocate when they need their place decorated. Wallpapering can be a tricky and messy process if attempting this on your own – or with a friend.

At this stage, there's still a little more preparation to do – sealing the surface. It would be a pointless exercise to paper over an unstable surface – damp or dusty surfaces could prove to be a problem. Any problems with dampness should be dealt with at the source (you will need to call in the professionals for this), but an old stain can be treated with a coat of proprietary damp sealant. New plaster or plasterboard must be given a coat of size (a stabilizing compound that is applied to seal porous surfaces before hanging wallpaper) or diluted wallpaper paste before papering. When dry, this seals the wall surface. Dusty surfaces need to be stabilized as well. A solution of PVA and water would be appropriate in this case. Refer to the manufacturer's instructions for the correct ratio.

1 The first thing you need to do is establish a starting point. In general you should start hanging wallpaper close to a corner, or if the paper has an obvious pattern, the first drop should be centred over a fireplace or other focal point in the room so that the pattern is symmetrical. If this is your first attempt at wallpapering, start near one end of the longest wall that has no obstructions, and with a bit of luck, you'll be an expert by the time you reach the other side.

2 Measure the wallpaper width minus 15 mm (⅝ in) away from the corner (this is to enable the drop of paper next to the corner to overlap the adjacent wall a little to disguise any gaps). Either get on the ladder with a hammer and a nail and hang the plumb line at this point to mark the true vertical, or use a spirit level and a pencil. This starting point is important, so make sure you get it straight.

* WOMEN HANGING THE WALLPAPER! THEY'LL WANT THE VOTE NEXT!

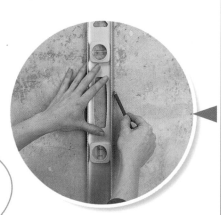

TIP

Don't panic if you see air bubbles when the paper is dry. Simply slit across the bubble carefully with a craft knife and then, using an artist's paintbrush, apply a little paste through the slit to the back of the paper. Smooth over with a brush or damp sponge.

3 Cut the first drop of paper, allowing about 5 cm (2 in) at the top and bottom edges. You can cut a few lengths of paper using this one as a guide and lay them all face down on the wallpapering table. This is when you need the string. Tie a length of string between the legs at one end of the wallpapering table. Tuck the ends of the paper behind it to stop the paper rolling up as you paste. Mix up the paste as directed on the container and apply it to the paper with a paste brush. Shift the paper up the table so that you can paste the entire length.

4 Fold the pasted paper up loosely in a concertina fold. Lift it off the table and set it aside for a few minutes for the paste to soak in. By the time you've pasted the second piece, the first will be ready to hang.

5 Time to hang the first piece. You'll need to get on the ladder for this. The best way is to position yourself on the ladder, then get your friend to hand you the pasted paper. Place the top edge of the paper along the coving, overlapping it a little and making sure that the edge is aligned with the marked vertical. Release the folds so that they unfold slowly. Use the wallpapering brush to smooth the paper out from the centre to the edges all the way down. This is to eliminate air bubbles and ensure good adhesion.

6 To trim the top and bottom edges, run the tip of your scissors gently along the angle between the wall and the coving. Then do the same with the angle between the wall and the skirting board. This is to make a crease as a cutting line. Gently pull the paper away from the wall, trim along the crease and then brush it back into position.

Plumb line

7 Hang the next piece of wallpaper in exactly the same way, taking care to make a good butt join with the first piece. Slide the paper gently into position if you need to match patterns. When you reach the opposite corner, cut the width of the paper so that it overlaps the adjacent wall by about 15 mm (⅝ in). Secure the edges and seams with a seam roller if necessary.

Going round the bend

CORNERS

All rooms will have corners – of this you can be absolutely sure. External and internal corners shouldn't be a problem if they are dealt with in a methodical way. The important thing to remember is that the corner may not be vertical. For example, if you try to paper around a corner and line the next piece up against it, you'll probably find that it's slightly uneven. By the time the room is complete, you'll be in a terrible mess. Even a few millimetres can make quite a difference. It may be a good idea to check your walls first. If they are really off true, you should avoid bold vertically striped wallpaper because it may look a bit strange in the corners.

For subsequent walls, repeat the process with the plumb line so that all the paper on each wall is true to the vertical and the corners will compensate for any unevenness.

INTERNAL CORNERS

1 Go back to the corner that you started in, the one where you measured the width of the paper minus 15 mm (⅝ in) from the wall. Hang a piece here and carefully brush the overlap into the corner. If the edge of the overlap does not lie flat or is wrinkled in places, make small cuts from the edge towards the corner. These are called release cuts and will compensate for any unevenness. Smooth the edge flat with a wallpapering brush.

2 Measure the width of the wallpaper from the corner outwards, then mark with a pencil. Take a spirit level and use it to mark a new plumb line. You can also use a special plumb line suspended from the ceiling. It is important to get this line absolutely straight; small discrepancies at ceiling level can turn into big discrepancies at skirting board level. If your wallpaper has a large or bold pattern, it will look obvious if the pieces have not been hung straight.

3 Paste and hang a new piece aligned to the plumb line. The edge will overlap the one that turned the corner, so there will be no gaps. If you are using vinyl paper, you may need a stronger adhesive to fix the overlap edge.

✳ NEXT TIME YOU'RE BACKED INTO A TIGHT CORNER, YOU'LL KNOW WHAT TO DO!

TIP

For an external corner, if the paper overlap doesn't lie flat, make small cuts along the edge, known as release cuts, to even things out.

EXTERNAL CORNERS

1 When you reach an external corner, hang the paper in exactly the same way, but this time tear the edge by hand so that it overlaps the corner by about 2.5 cm (1 in). The reason for tearing by hand is that it leaves a ragged edge, which is likely to make a less noticeable ridge under the edge of the subsequent piece.

✳ CAN ALSO BE USED ON YOUR HIPS!

need
✳ scissors
✳ craft knife
✳ screwdriver

2 Measure and mark a new plumb line, the width of the wallpaper plus about 15 mm (⅝ in). Cut the next piece of paper to size, taking care to match any patterns. Then paste and hang using the plumb line as a guide. Smooth the piece in place with the wallpapering brush. The overlapping edge will probably need a little more than a quick brushing over to make it adhere properly. Use the seam roller up and down the edge to make sure that the edges are firmly stuck down. If the edges lift slightly when the paste is dry, apply a little more paste with a small artist's brush.

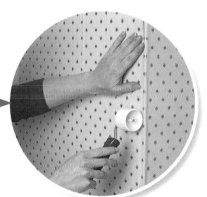

PAPERING AROUND LIGHT SWITCHES

Most rooms have light switches too – it's just one of those inevitable things. The next two steps will show you how to paper around them successfully.

1 Paper over the light switch and use the craft knife to make diagonal cuts in the paper from the centre of the light switch towards each corner. Use your brush to ease the paper to fit around the edge. Trim the triangular tabs to about 15 mm (⅝ in) of the edge.

CAUTION
Switch off the electricity at the mains before you do this. But make sure that it is light enough in the room to see what you're doing.

2 Loosen the screws of the switch faceplate and use the brush to tuck the trimmed edges behind it. Screw the faceplate securely into position again. Don't turn the electricity back on until the paste is dry. Electricity and moisture don't mix.

MORE FASCINATING WALLPAPER FACTS

Some rooms can turn out to be a veritable obstacle course of awkward nooks and crannies, such as window recesses, air vents and so on. Similar wallpapering methods are employed to deal with all these little problems.

First, you need to start hanging the wallpaper on the largest flat expanse of wall. By the time you reach the other corner or a little obstacle, you'll probably find you have a fairly firm grip of the general principles involved and will be more confident about handling the obstacle.

Remember what you learned about external and internal corners and light switches on the previous pages? Just combine all those bits of information and you'll soon be wallpapering in and out of window recesses and around doors and air vents like a demon. It's simply a case of application. Everyone's nooks and crannies will be slightly different, so you just have to modify these basic methods to suit your needs.

This isn't difficult to do – a little complicated maybe, but certainly not rocket science. All you need is a little patience – and all the tools for the job of course! The fact is, unless you paper right over all the doors and windows in the room, you're just going to have to get to grips with it, period!

TIP
If the wallcovering has a repeat pattern, the general rule is to increase the length of each drop by the measurement of the pattern. This allows plenty of space for pattern matching. If the pattern is very large, it is best to hang one piece at a time, cutting the next to match.

AIR VENTS

1 Paste and hang the wallpaper so that it overlaps the edge of the vent. Make a diagonal cut from the edge of the paper towards the corner of the vent, forming triangular flaps. Brush the paper gently into the angle between the vent and the wall. Brush the remaining paper down the side of the vent towards the skirting board, cut towards the corner as before, then brush the paper into the angle between wall and skirting board. Use the wallpaper brush to press the paper firmly into the edge of the vent to ensure that it is firmly stuck.

2 Use a craft knife to trim along the crease. Trim the edge next to the skirting board using scissors in the usual way. Hang the next piece of wallpaper, matching any patterns carefully; then cut and trim to fit around the other half of the vent in exactly the same way. Trim top and bottom edges as before. If the vent falls in the centre of a piece of paper, deal with it in the same way as the light switch.

WINDOW RECESSES

AND FINALLY FOR THE ADVENTUROUS!

Window recesses (indeed any recesses) are a nasty combination obstacle course of internal and external corners. There will be lots of trimming, tearing, pasting, brushing and patching of paper to fill gaps. Then there is the pattern-matching thrown in just for good measure. Don't get discouraged – it really is a simple and logical process.

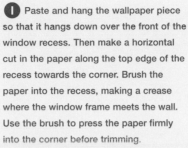

1 Paste and hang the wallpaper piece so that it hangs down over the front of the window recess. Then make a horizontal cut in the paper along the top edge of the recess towards the corner. Brush the paper into the recess, making a crease where the window frame meets the wall. Use the brush to press the paper firmly into the corner before trimming.

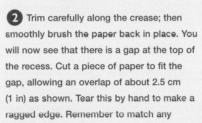

2 Trim carefully along the crease; then smoothly brush the paper back in place. You will now see that there is a gap at the top of the recess. Cut a piece of paper to fit the gap, allowing an overlap of about 2.5 cm (1 in) as shown. Tear this by hand to make a ragged edge. Remember to match any patterns when you cut the extra piece.

3 Carefully lift up the edge of the piece next to the window. Paste and position the extra piece, then smooth the first piece over it to cover the ragged edge. Hang straight pieces that tuck under to cover the top of the recess until you reach the other side of the window; then fill in the corner part in exactly the same way as the first.

✳ REMEMBER, GIRLS, WEARING SPIKE HEELS ON A LADDER IS NOT A GOOD IDEA!

At last, here comes the glamour part!

PAINTING

Paint manufacturing companies bring out new shades and products all the time to satisfy the ever-increasing demand, so you should have no problems finding the colour you want. Most DIY stores have special paint-mixing systems – you just take an example of the colour you want, the store will match it and mix it in the quantity you need.

WHAT IS PAINT?

Paint basically consists of three parts: binder, pigment and carrier. The binder helps the paint stick to the surface and holds everything together. The pigment is the part that gives it colour, usually a white base with added dyes, and covers the surface underneath. Finally, the water- or oil-based carrier makes the paint flow smoothly and evaporates as it dries.

WATER- OR OIL-BASED?

To make life simple I'm going to put paints into two categories: oil-based (such as gloss paints) and water-based or emulsion paint. Water-based paints are generally easy to apply and less likely to dry leaving brush marks. Oil-based paints tend to take a long time to dry and give off an unpleasant smell, although most manufacturers now produce low-odour paints. This type of paint is more durable than emulsion paint; it is used for woodwork and metal fixtures. Oil-based

PAINTING IS THE MOST POPULAR CHOICE FOR HOME DECORATING. IT'S EASY, QUICK AND THERE IS A WHOLE SPECTRUM OF PAINT COLOURS TO CHOOSE FROM.

✱ OF COURSE IT'S ALL A QUESTION OF TASTE... I ALWAYS THINK YOU CAN'T GO WRONG WITH BLACK.

paints will require a special primer or undercoat for good results, while water-based paints do not – simply apply two or three coats for even coverage.

MATT OR SHINY?

The lower the ratio of pigment to binder, the shinier the finish. Matt-finish paints contain more pigment particles, so, when dry, the surface is a little rough and doesn't reflect light like a smooth surface would. Satinwood, eggshell, gloss or high-gloss paints have a lower ratio of pigment, dry more smoothly and reflect more light. Gloss, high-gloss or eggshell finishes are usually used for woodwork or metal. Satinwood or matt-finish paints are used for walls and ceilings. Matt finishes will cover up imperfections while gloss will accentuate any lumps or bumps.

DIFFERENT PAINT

Textured paint

Textured paint is very thick and gives a three-dimensional textured appearance. It can be applied with a brush or roller to create effects such as rough plaster or sandstone. This paint is great if your walls are less than perfect and it can cover a multitude of sins! Textured paints are available for exterior masonry and concrete.

Gloss paint

This is used for wood or metal and dries to a nice glossy finish. This type of paint is oil-based and takes quite a long time to dry, but is very durable and easy to wipe clean. Remember, primer and undercoat should be used to achieve good results when using a gloss paint system.

Melamine paint

Ingenious paint designed to cover up that ubiquitous melamine! Use it together with a special primer for a fabulous quick fix.

Tile gloss

This paint is specially formulated for use on ceramic tiles. Tile gloss must be used with a tile primer for good results.

Metallic paint

Metallic-effect paints are really popular these days, both in brush-on and in spray form. Various products are available for decorating walls, wood, metals and plastics.

Water-based emulsion paint

Emulsion is probably the paint you'll be most familiar with. This water-based paint gives good coverage, is fairly quick-drying and is easy to use. Available in matt or satinwood finish.

Now for some fun

BASIC PAINTING

There is definitely a right way and a wrong way to do everything, and this is certainly true of painting. Applying the paint in a haphazard fashion will not achieve good results, so it's best to have a plan. As I've suggested already, you should strip everything that needs stripping and complete all the preparation beforehand, then take care of painting the woodwork and ceilings first.

need
* paint
* paintbrush
* masking tape
* primer

PAINTING WOODWORK

1 Prepare the woodwork before you begin painting. Strip off the old paint (*see page 133*), sand smooth and clean the woodwork as necessary.

2 If the woodwork is already bare, apply a coat of a suitable primer before painting. The primer will provide a base for the top coat. The notes on the paint tin will tell you which wood primer to use.

3 For narrow areas such as skirting boards, brush the paint in the direction of the grain. For larger areas such as wooden door panels, brush first with the grain, then join the strokes by brushing across the grain. Rebrush again with the grain, allow to dry, then recoat.

4 When you begin to apply paint to the wall areas, mask off the edges of pre-painted skirting boards or other woodwork, then use a small brush to 'cut in'. Apply paint around all the edges in this way. Then apply paint to larger areas using a large brush, roller or pad.

✻ WAAH! I WANT TO PAINT TOO!

TIP
A word about paint. In general, it is false economy to buy cheap paint. Coverage is far better with a good-quality paint. Cheap emulsion will need three or four coats, instead of the usual two coats. Time-consuming and tedious!

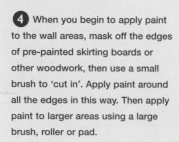

PAINTING WINDOWS AND DOORS

PAINTING WINDOWS

Painting sash windows can be a little tricky, but there is a system. The two parts of the window have to be free to slide up and down, so be careful not to stick them together with paint (*see diagrams right*).

Casement windows have at least one hinged pane that opens outwards. If you prop the window open while you paint, there is no possibility of it being stuck in the closed position.

When painting windows, try not to get spatters on the glass. Mask off the edge of the pane all around with masking tape.

1 For sash windows, first begin by raising the lower sash and lowering the top one. Next paint the lower horizontal rail of the top sash (1) and any other vertical areas (2) you can reach easily.

2 Close the sashes, but leave a slight gap. Paint the remaining areas of the top sash (3) and then the bottom sash (4). Now paint the outer frame (5). When the paint is dry, paint the inner runners).

1 For a casement window, with the window open, paint the back edge of the sash (1). Next paint the face of the sash (2), followed by the inner side of the frame (3). Finally paint the sill (4).

PAINTING PANELLED DOORS

Painting panelled doors requires a great deal of care and attention. Each part of a wooden panelled door must be painted in order; it makes it easier to do and gives a better finished result. See the diagram for the order of work; each part is numbered. For good results, you need to work quickly, always paint in the same direction of the grain and remember to remove any loose bristles while the paint is wet. To avoid unsightly paint build-up at the edges, don't wait for each section to dry before proceeding to the next.

BASIC

1 Begin with the panel edges (1); then paint the large vertical panels (2). Next paint the central vertical part (3), and then the horizontal cross rails across the top, centre and base (4). To complete the door, paint the outer verticals on each side (5).

ADVANCED

1 For a very professional finish, follow the order and direction of brushing on the diagram. This more advanced method (6–12) ensures that any wet edges are feathered in to avoid visible joins between sections on the finished door.

Slap it on...
PAINTING WALLS

After the skirting board and any other woodwork has been done, you can proceed to the main paint job. There are various ways to apply paint: paintbrush, roller and paint pad. For small, awkward areas or for painting straight lines next to door frames or windows, it is best to use a small brush (*see page 148 –* painting woodwork), but for larger areas you can use a large brush, roller or pad.

BRUSH ON

1 Using a brush: when using a brush, dip about one third of the bristle length into the paint – it gets a little messy if you dip right up to the handle. Press the bristles lightly against the rim of the tin to remove excess, then apply to the wall. Use random crisscross strokes over small areas at a time. Apply a second or subsequent coats only when the previous one is completely dry.

2 Using a roller: paint rollers come in varying sizes and are a quick and easy way of covering large areas. You will need a roller tray. Pour some paint into the well of the roller tray, then pass the roller through the paint.

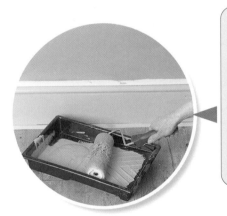

ROLL ON

3 Roll the sleeve up and down the sloping ridged part of the tray to cover it evenly with paint. This is known as 'loading' the roller. If the roller is not fully and evenly loaded, the result will be a patchy finish on the wall. Apply paint to the wall in an up, down and side-to-side motion to ensure even coverage.

PAD IT OUT

4 Paint pads are a different method of paint application. They are quick and easy to use and they also give good coverage. Pour paint into a roller tray or paint pad trough, coat the foam pad with paint, then apply to the wall in random crisscross strokes.

✳ IF THAT'S YOUR TASTE IN TIES, YOU'D BETTER LEAVE CHOOSING THE PAINT TO ME!

PAINTING CEILINGS

So you're standing there looking upwards, wondering how you're going to stand on the ladder, paint a bit, climb off the ladder, move the ladder, get on the ladder and so on. Don't panic – you never use a ladder to paint a ceiling (unless it really is lofty, and then you'd have to hire the proper equipment). There is a simple answer to the how-to-paint-the-ceiling dilemma – a paint roller handle extension. This is a simple device that fits over the handle of your roller and is adjustable to fit the distance between you and the ceiling. The next vital piece of equipment to remember is a shower cap! This may sound and look a little amusing, but it's much better than getting paint splashes all over your new hairdo.

need

* paint roller with extension handle
* roller tray
* shower cap

Avoid a bad hair day. Put on a shower cap before painting the ceiling.

✽ HEY, YOU CAN PAINT MY CEILINGS ANY DAY!

PAINTING BEHIND RADIATORS

Almost all houses and flats have radiators or wall-mounted heaters of some sort. It isn't very professional to leave an area unpainted just because you can't reach behind the radiator. It is possible to poke your paintbrush behind the radiator and hope for the best, but really, all you need to do is buy a mini roller with an extra-long handle. This enables you to reach right down behind the fixture quite easily.

☞ Load the paint roller with paint and away you go. Try not to get any paint on the edges of the radiator. If you do, just wipe clean with a damp cloth before it dries, or you'll have to scrub it off. If it does not interfere with the roller, you could lay a dust sheet over the radiator to protect it before you begin.

Now it's time to get really arty

SPECIAL EFFECTS

This is the colourful, creative part. Decorative paint effects make a change from flat colour. You can choose any type of effect you like – make it subtle or bold, calm or loud. Most paint effects don't require any special equipment, but it does take a little practice to achieve the result you want. It's a good idea to try out the effect on an unobtrusive area – behind a door, for example – until you feel confident to do the rest of the room. All these paint effects start with a base coat in the colour of your choice, then the effect is applied on top.

* I'M READY FOR MY CLOSE-UP NOW...

COLOUR WASHING 85

This involves applying a watery paint mixture over a pale base colour. The result is a lovely delicate mottled effect. You can use one, two or more colours or shades of the same colour to achieve depth.

need

* emulsion paint in a pale and darker shade of your chosen colour
* kitchen cloth (or sponge or large brush)
* paint kettle

❶ Apply base coat and allow to dry. Next mix up a watery solution of the palest colour in a paint kettle. Dip the kitchen cloth in the solution, then wipe it over the entire surface in random circular sweeps. If this looks too pale, you can always add more, but it's difficult to take it away.

❷ When the first coat is dry, repeat the procedure using a darker colour wash. This colour-wash effect can be achieved in the same way using a natural sponge or a large paintbrush.

❸ The result is a very delicate cloudy effect. This works beautifully using a combination of pale colours, but it can also be very striking when done with bolder, more contrasting shades.

SPONGING  86

Sponging results in a more defined but still subtle mottled effect. It is important that you use a natural sponge for this, not a synthetic one. Natural sponges have irregular holes and therefore produce a more random imprint. There are two ways to do this: sponging *on* for a clearer mottled pattern or sponging *off* for a softer finished effect. Try a small patch of both in an unobtrusive place to help you decide which looks best.

GET IT OFF

1 To begin, apply one or two coats of your chosen base colour, allowing each coat to dry before applying the next. You can do this using a brush, pad or roller. It is important to allow the base coat to dry completely before starting any paint effects. If the base is damp when the decorative layer is applied, the two colours will smudge and blend together, then the full effect will be lost.

2 To sponge off, use a brush to apply a solution of 1 part paint to 4 parts scumble glaze over the base coat using a brush. The scumble glaze prevents the paint from drying too quickly, allowing a little extra time to adjust the effect if necessary. Using the sponge, dab the wet glaze, lifting off the paint to reveal the base coat. Rinse the sponge when it becomes clogged with paint.

GET IT ON

3 To sponge on, simply dip a slightly damp sponge into the paint, then dab off any excess on a piece of paper towel. Press the sponge lightly on the wall, working in random arcs so that you don't create any regular patterns. It helps too to turn the sponge around in your hand every now and then to vary the pattern a little. Rinse the sponge with clean water occasionally to prevent the holes from clogging up with paint.

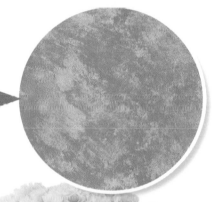

need

* natural sponge
* emulsion paint
* scumble glaze
* paintbrush
* paint kettle

RAGGING AND RAG ROLLING

87

Ragging and rag rolling result in a random pattern, but the effect is less subtle than sponging. All you need is a piece of lint-free cloth. The cloth is scrunched up and used to lift off a glaze or to apply colour to the base coat. The creases in the cloth create the pleasing characteristic patterns.

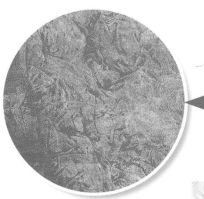

need

* lint-free cloth
* emulsion paint
* scumble glaze
* paintbrush
* paint kettle

RAGGING

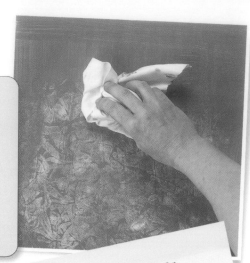

1 Apply a solution of 1 part paint and 4 parts scumble glaze to the dry base coat. Take a piece of lint-free cloth and crumple it up into an irregular pad. Simply use the pad in a similar way to sponging off (*see page 153*). Dab the cloth firmly onto the wet surface of the glaze, lifting it off to reveal the base coat underneath.

2 Recrumple the cloth occasionally, so that the pattern varies slightly over the expanse of the wall. If you do not do this, there is a danger of creating regular blotches or stripes, which will spoil the overall effect.

A plain wall is an unloved wall... All it needs is just a little rag and roll.

RAG ROLLING

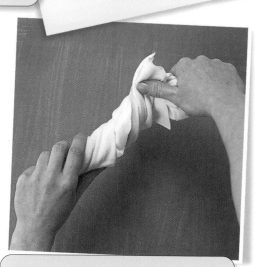

1 The rag rolling method is similar to sponging off. Apply a solution of 1 part paint to 4 parts scumble glaze over the base coat. Take a clean piece of lint-free cloth and roll it up into a loose sausage shape. Have a few clean cloths ready so that you can replace it when necessary.

2 While the glaze is still wet, roll the rag over the surface from the bottom towards the ceiling to create the patterned effect. Work along the surface in overlapping bands so as not to form noticeable stripes.

3 Rinse out the cloth with water, then reroll when it becomes clogged with paint. You can also lightly brush over the finished wall with a large, soft paintbrush in order to achieve a more subtle effect.

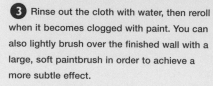

BAGGING 88

This is something that you can do with all those plastic carrier bags from your trips to the supermarket. Bagging creates a strong random pattern and is a lot of fun to do. You can look at this as a form of double recycling – after you've used the bag for the shopping, you use it again for the decorative paint effects. When you've finished using it for decorating, rinse it out and then you can take it to the recycling point.

need
* plastic bag
* cloth
* emulsion paint
* scumble glaze
* paintbrush
* paint kettle

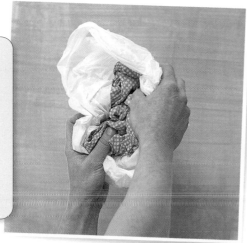

1 Mix up a solution of watered-down emulsion paint with scumble glaze in a paint kettle, then apply over the base coat. Wrap the cloth inside a plastic bag, then crumple it up. The cloth is used to give the plastic bag some volume, forming large, firm, irregular folds. The bag itself would be too flimsy to create a substantial decorative pattern on its own.

2 Take the crumpled bag in both hands and scrunch it all over the wall in a light kneading motion, lifting off the glaze to reveal the base coat underneath. Discard the bag and use a new one when it becomes too coated with paint. This may feel a little slippery, because the bag slides over the surface of the wet glaze. Remember to lift the bag away from the surface as you move from section to section or the effect could smear.

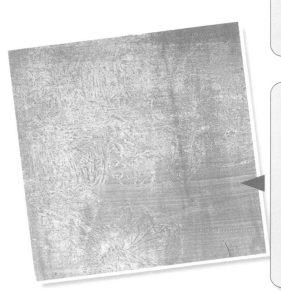

3 The result is a mottled effect with a difference; the patterns created are random, yet quite angular at the same time. This effect is very similar to rag rolling but does not have the softness that using a cloth produces. The same precautions should be taken to avoid creating obvious stripes or blotches which will spoil the final appearance.

* I NEED MORE BAGS – I'LL JUST HAVE TO GO SHOPPING AGAIN. WHAT A DRAG!

Fun to be had with a flogger brush and roller

DRY BRUSHING AND DRAGGING

89

This paint effect creates fine lines almost like grain lines, either horizontally or vertically over the base coat. The only piece of special equipment you'll need is a brush with long, soft floppy bristles called a 'flogger', or dragging, brush. This enables you to apply the paint in fine even lines.

need

* emulsion paint
* scumble glaze
* paint kettle
* flogger brush
* cloth
* paintbrush

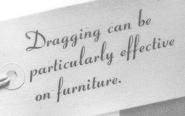

Dragging can be particularly effective on furniture.

DRY BRUSHING

1 Apply two or more coats of base colour to complement or contrast with the top coat. When this is dry, apply a top coat of emulsion paint mixed with scumble glaze.

2 Dip a flogger brush in a little emulsion paint. Wipe off excess paint on a clean cloth, then drag the brush lightly over the surface. This creates a softer dragging effect. When painting furniture, a dry brush can be used to add a little mottled colour to edges and mouldings for a decorative aged effect.

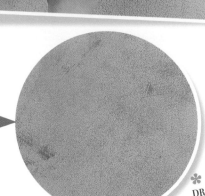

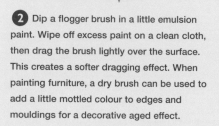

DRAGGING

1 Again, choose a base colour to complement or contrast with the top coat. Apply two or more coats and let dry.

2 Apply a coat of emulsion mixed with scumble glaze using a regular paintbrush. While it is wet, drag the dry flogger brush firmly through the glaze, creating fine lines. Use a cloth to wipe off any excess paint/glaze solution.

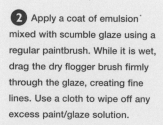

* IT'S JUST A DRAG, OKAY!

USING ROLLERS

90

Rollers can also be used to create unusual paint effects. The sleeves can be made from natural or synthetic fibres and can be smooth or have a long or short pile. Long pile is best for painting masonry or highly textured surfaces, while a shorter pile is best for flatter surfaces to be used with emulsion paints. If you are using gloss or varnish, a smooth foam roller is ideal; this will produce a professional finish. If the paint is oil-based, the sleeves can be difficult to clean, but the foam sleeves are relatively inexpensive and can be thrown away after use.

USE ROLLERS FOR A SMOOTH FLAT FINISH OR FOR WILDER TEXTURED PATTERNS.

Short-pile paint roller

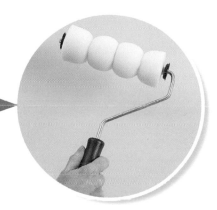

1 Rollers are great for applying paint quickly to large areas. Use a regular roller for the base colour, then a textured one to create a surface pattern. Subtle or bold – the choice is yours.

2 You can customize a foam roller sleeve by tying pieces of string tightly around it to create an irregular striped pattern. You can even tie the string in a random fashion for an even more irregular effect. For a very strong irregular pattern, try cutting uneven chunks out of a thick foam roller sleeve.

3 Small rollers are available with specially textured sleeves for home-decorating projects. Use in the same way as any other roller; just coat with paint and roll. This one looks like an animal print. The base of the foam sleeve has cut-out shapes in a softer foam glued to it. As you roll the sleeve through the paint, only the raised areas are coated, thus creating the pattern when applied to the wall. Textured rollers can also be made from moulded rubber to combine texture with pattern.

* NO, NOT THOSE KIND OF ROLLERS!

This is where you can get really imaginative

✳ GO FORTH AND STENCIL, GIRLS!

STENCILLING 91

A very popular craft, stencilling is a way of creating repeat patterns or single motifs on a multitude of surfaces, including paper and fabric. Since the development of thermo-hardening paints, permanent images can be created on glass and ceramics.

The principles of stencilling are very simple. The image is cut out of a piece of very thin waxed cardboard or acetate film. The stencil is then placed on the surface and usually held in place with strips of masking tape or a spray adhesive. Paint is applied sparingly through the stencil, using a fat, stubby round brush and a dabbing or stippling action. The secret is to apply as little paint as possible in order to create a soft shaded effect. Many colours may be used together if you use a different brush for each colour. Machine-made stencils can be bought from art and craft shops, or you can easily make your own.

The most important thing to remember is to use specially formulated stencil paint or quick-drying acrylic paint. There is a danger of smudging the design if paints with a long drying time are used – this is something to be careful of if attempting to stencil on glass or ceramics. You must also take care not to move the stencil once the painting has begun, because this will cause a double-image effect. Start with a simple one- or two-colour motif, then progress to more complicated designs.

❶ Trace the simple daisy motif given on page 182 or design one of your own. When designing, remember to make sure that all the pieces have a margin of acetate, however narrow, around each one; otherwise you will find that your motif will drop out of the acetate and the design will be spoiled. Check the design carefully before you begin cutting.

START WITH A SIMPLE DESIGN.

need
* **paper and pencil**
* **stencil acetate**
* **scalpel**
* **masking tape or spray adhesive**
* **cutting board or safe surface**
* **plumb line**
* **large square of cardboard**
* **stencil brushes**
* **acrylic artist's paint: white, blue, green**
* **paint palette or flat plate**
* **kitchen towel**

TIP
If you choose to use this motif continuously as a border or vertical stripe, you will need to mark a centre line.

❷ Cut a piece of acetate stencil film to size and fix it to the tracing with masking tape. Place on a cutting board or safe surface, then carefully cut out each piece using a scalpel. Remove the finished stencil and then repeat the process to make another. Use this one in reverse. If you accidentally make a slip and cut through more that you meant to, simply patch it with a piece of clear tape for a quick repair.

3 You must now mark out the position of each motif on the wall if it is to be an all-over pattern. This is a very simple way of marking out a diagonal grid on a large scale. First hang a plumb line from the top of the wall in a central position. Cut a square of cardboard the size of the spacings between the motifs (i.e. the centre of the motif will be at each corner of the square marker). The spacer can be as large or as small as you like, depending on the size of the motif.

4 Place the marker behind the plumb line so the string hangs down the centre. Mark each corner with a faint pencil cross. Move the marker down so that the topmost corner lines up with the previous lower cross and then mark each corner again. Continue in this way so the entire wall is marked into a grid.

HEY, WHY NOT COORDINATE YOUR OUTFITS WITH YOUR DECOR! ON SECOND THOUGHTS...

5 Attach the stencil to the wall with masking tape. Make sure that the centre of the motif lies over each pencil cross. Pour a little blue paint onto the palette, then take up a small amount of paint on a stencil brush. Dab off the excess on a piece of kitchen towel, then apply the paint through the stencil using a dabbing or stippling motion just at the centre and tips of each petal.

Stencilling brushes

FANTASTIC FOR APPLYING MAKE-UP TOO!

Keep on stencilling

6 Using a different brush, apply white paint to the remainder of the petals in the same way. Apply less pressure with the brush when you reach the petal tips and towards the centre to create a softer, less solid appearance. The idea is to blend the colour gently and not to cover the first colour completely.

7 For the stem and the leaf, use a smaller stencil brush and green paint. Apply paint to the leaf quite solidly at the base and tip, but more sparingly towards the centre to give a slightly curved effect. The same method is used to blend colours together; just apply the paint sparingly at the point where the two colours meet. Use the mirror-image daisy stencil alternately to form a flowing pattern all over the wall.

Move on to more complicated designs when you feel more confident.

8 Use the same colours for all the daisy motifs, or use a contrasting scheme for the mirror-image motifs. If you prefer a simpler approach, use one colour only, and use subtle shading to create depth. Metallic paints can also be used to great effect. Choose a colour close to the base shade; the motif will be picked out when light is reflected from it.

STAMPING

Stamps are flat blocks of wood or foam that have a raised motif glued to or cut out of one side. The image is coated with paint, using a small foam roller, and transferred to a surface by simply pressing the stamp firmly to it. Recoat with paint and repeat for multiple images. All-over patterns on walls should be marked out in the same way as for stencils. For a border or stripe, either mark the horizontal or vertical line with faint pencil marks or stick a strip of masking tape to the wall, then align the top edge of the stamp with it. Use a spirit level to ensure that horizontal and vertical lines are true.

need

* stamp
* pencil or masking tape
* acrylic paint
* small roller
* paint palette or flat plate

1 To stamp a border pattern, decide on the height, then mark the walls with pencil or low-tack masking tape. Pour a little paint onto a flat plate or paint palette, and pass the roller a few times through the paint to coat it completely. Hold the stamp in one hand and then pass the roller over the motif, coating it evenly. For a regular all-over pattern, use a cardboard spacer in the same way as for the stencil. For random patterns, just press the stamp over the wall, rotating it a quarter turn each time so that each is at a different angle.

2 Press the motif side of the stamp firmly to the wall in the correct position. Lift the stamp off squarely; do not move the stamp while it is on the surface or the image will be blurred. Recoat the stamp with paint, then repeat to complete the border or all-over design. You can probably make two impressions with one coating, but any more than that and the image will begin to fade.

* DARLING, YOU'LL NEVER GUESS WHAT I'VE STAMPED ON THE BEDROOM WALLS!

FOR STENCILS AND STAMPS

☞ As mentioned previously, since the development of thermo-hardening paints, it is possible to make permanent decorative images on ceramic surfaces. This gives a great deal of scope for decorating wall tiles.

☞ Why not use stencils to create a pattern or border on the floor of your room, on the wooden floorboards or even on vinyl flooring? You must use paint suitable for floors, or vinyl, because regular paint will wear off.

☞ Paint your staircase one colour and try out decorative stencil patterns along the edges of each step.

BRING A TOUCH OF GLAM TO YOUR KITCHEN!

Revamping the refrigerator 93

STAINLESS STEEL refrigerators are very trendy, but the price tag can be slightly scary. Fortunately, the same look can be achieved for a fraction of the price by using metallic spray paint. Spray paint is perfect for use on metal, enamel or even plastic surfaces. The advantage is that there are no brush marks to be seen. There are, however, a few major considerations. First, spray paint has a tendency to go everywhere. Preparation plays a very important part too. The shiny metal or enamel surface must be properly rubbed down prior to painting or the paint will not adhere properly and the finish will be ruined. Metallic spray paints are available in smooth and hammered finishes.

1 Cover all surfaces with dust sheets to protect them from the 'fall-out'. The fine paint spray hangs in the air for a while, then settles on anything and everything. Ideally, you should take the object to be sprayed outdoors, cover the immediate surroundings with a dust sheet, then spray. If you don't have a garden, you will have to spray indoors, so be careful to cover everything, open all the windows and always wear a mask.

need

* refrigerator
* dust sheets
* face mask
* masking tape
* sheets of newspaper or brown paper
* wet and dry sandpaper sheets or sanding sponge suitable for metal
* cloth
* spray paint (2 cans will cover an average-sized refrigerator)
* large sheet of stencil acetate
* scalpel
* hammered-finish metallic spray paint

2 Mask off all the rubber seals around the refrigerator and on the door. Tape a sheet of newspaper or brown paper over the front of the refrigerator so that the interior doesn't accidentally get sprayed silver. Now the hard work starts. Use dampened wet and dry sandpaper or a sanding pad for metal. Rub all the surfaces to be painted thoroughly – all the shine must be removed in order to create a good key for painting. Then wipe down all surfaces with a damp cloth to remove all the dust.

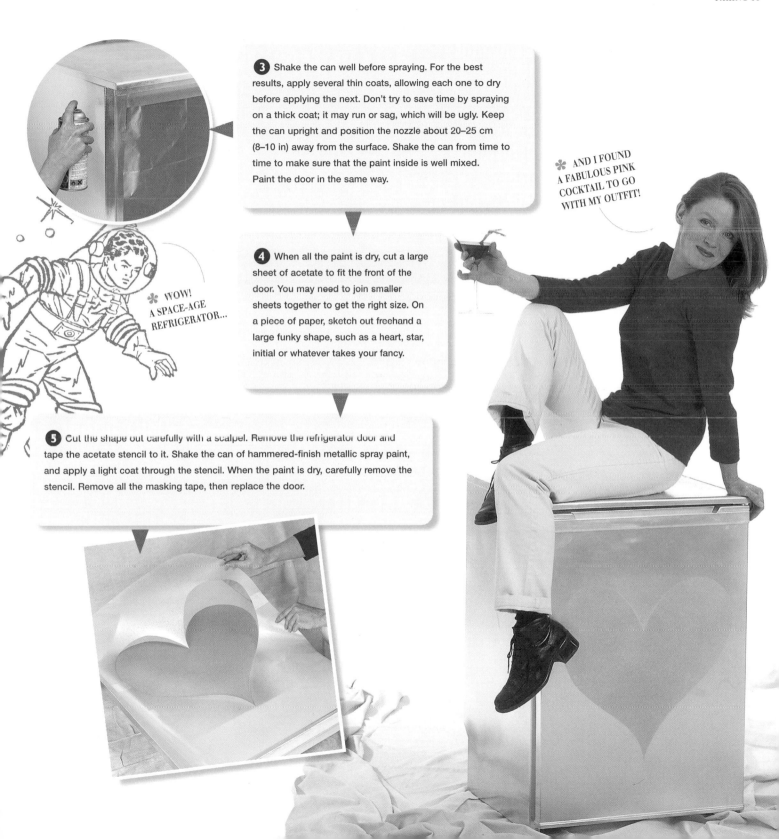

3 Shake the can well before spraying. For the best results, apply several thin coats, allowing each one to dry before applying the next. Don't try to save time by spraying on a thick coat; it may run or sag, which will be ugly. Keep the can upright and position the nozzle about 20–25 cm (8–10 in) away from the surface. Shake the can from time to time to make sure that the paint inside is well mixed. Paint the door in the same way.

✱ AND I FOUND A FABULOUS PINK COCKTAIL TO GO WITH MY OUTFIT!

4 When all the paint is dry, cut a large sheet of acetate to fit the front of the door. You may need to join smaller sheets together to get the right size. On a piece of paper, sketch out freehand a large funky shape, such as a heart, star, initial or whatever takes your fancy.

✱ WOW! A SPACE-AGE REFRIGERATOR...

5 Cut the shape out carefully with a scalpel. Remove the refrigerator door and tape the acetate stencil to it. Shake the can of hammered-finish metallic spray paint, and apply a light coat through the stencil. When the paint is dry, carefully remove the stencil. Remove all the masking tape, then replace the door.

And now on to techniques for stripping furniture

STRIPPING FURNITURE 94

Furniture is far more complicated to strip than a flat surface because it is likely to have all sorts of oddly shaped corners, legs, panels and decorative parts. Stripping chairs can be really frustrating – you've stripped every possible surface, then you turn the chair around and spot a few more that you missed. The best way to do it is to turn the chair upside down first, and do the underside surfaces that are easily accessible. Then turn it upright and work on each side at a time, then the top flat surfaces and seat. That's the general idea with most furniture – you need to look at it from all angles. There are a few different methods you can try.

✱ STRIPPING IS AN ART FORM!

need
* hot air gun
* chemical paint stripper
* wire wool
* white spirit
* shave hook
* flexible filling knife
* old newspaper

HOT AIR

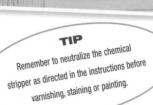

❶ A hot air gun is fantastic at removing paint from most surfaces. Just point the gun at the paint, pull the trigger and let the hot air melt and bubble the paint. You can then scrape the paint off easily using a flat scraper or a shaped shave hook. Use newspaper to wipe the melted paint residue from the scraping blade regularly. Remember that the melted paint is hot, so don't drop it on your skin. Be careful not to scorch the wood – this isn't a problem if you intend to paint, but it may show through a stained or varnished finish.

TIP
Remember to neutralize the chemical stripper as directed in the instructions before varnishing, staining or painting.

PEEL IT OFF

❷ Chemical stripper does the same job, but it will not leave marks on the surface, which is good for flat areas and also mouldings or delicate woods. Simply apply the stripper thickly, using a paintbrush, and leave it for the required amount of time. The paint will have bubbled up nicely and will be easy to scrape off. Again, keep plenty of sheets of newspaper handy for the paint residue.

SCRAPE IT

❸ A shaped shave hook is a handy tool when stripping furniture. The triangular blade is shaped differently on each side and has pointed corners, just right for gaining access to tiny spaces and mouldings.

Tired of that old furniture?
Short of cash? Give it a new look

PAINTING FURNITURE

95

The same principle applies to painting furniture. Devise a method and stick to it: paint the undersides first, then one side at a time. For wood furniture the general rule is to paint with the grain. As with most paint jobs, it is far better to apply a few thin coats than one thick one. You may think you are doing yourself a favour, but one heavy coat is a false economy of time. Use small brushes for easy access to detailed areas.

MELAMINE

Melamine is pretty hardwearing, cheap and easy to clean. If you're stuck with ugly melamine kitchen units, a garish chest of drawers or a dated wood-effect wardrobe, a quick paint revamp is a much cheaper option than getting rid of the whole thing. Use specially formulated melamine paint together with melamine primer.

need

* wet and dry sandpaper sheets or sponge sanding pad
* cloth
* roller tray
* paint roller
* melamine primer
* melamine paint
* screwdriver

1 Unscrew and remove all handles. Rub down all surfaces to be painted with dampened wet and dry sandpaper or a sanding pad. The preparation method is similar to that for metal. The shiny surface of the melamine must be roughened to create a good key for the primer. Pour some melamine primer into the roller tray, load the roller and then apply a thin, even coat to all prepared surfaces. Don't forget the edges and the undersides of the doors.

2 When the primer is completely dry, apply a coat of melamine paint in the colour of your choice. You may need to apply two coats for good even coverage. Allow each coat to dry before applying the next. Replace all door fixtures when the paint is dry – you could always add some new handles. Throw the old handles away or keep them to reuse for another project.

3 A fresh paint job really makes a difference, and it may be just the thing to keep you happy while you save up for the full kitchen renovation.

✳ AND I'VE STILL GOT TIME TO GIVE MY MAN SOMETHING SUBSTANTIAL FOR DINNER!

How to fake it like a professional

Here's another selection of paint effects for you to try. Most are more suitable for furniture or small decorative pieces, such as picture frames or storage boxes. Larger items, such as doors, can also benefit from some of these special effects, but on the whole they're not suitable for large expansive areas such as walls.

DISTRESSING 96

This is a really easy way to achieve an aged, almost antique, appearance. You can make a relatively cheap piece of junk-shop furniture look like a family heirloom. Prepare all surfaces to be painted as usual, then apply two or more coats of your base colour, allowing each coat to dry before applying the next.

Remember it's only the wood that you're meant to distress!

need
* sandpaper or coarse-grade wire wool
* wax shoe polish, dark or light tan colour
* soft cloth

1 Using sandpaper or a small pad of coarse wire wool, rub away the base coat at areas that are likely to have received wear and tear (over the centuries, if you're creating an heirloom): corners, edges, round drawer handles and edges of raised moulding. You need to remove the top layer of paint in these areas to reveal the wood underneath.

2 Wipe away any particles of paint and dust with a damp cloth. Now dab a little shoe polish on the cloth and work it into the sanded areas. This gives a slightly grimy, well-worn, aged appearance. This effect can be further enhanced by adding a slightly tinted varnish over the top.

3 Buff the finished article to a soft sheen with a soft cloth.

* AND HERE'S THE FINISHED EFFECT! SO DISTRESSING!

RESIST EFFECTS

97

This is another way to achieve a distressed, worn appearance. This works best if two contrasting-colour paints are used, or a dark and a light shade of the same colour. The idea is to apply a coat of beeswax to certain areas of the base coat. The top coat will not stick to the wax and is rubbed off to reveal the contrasting colour underneath. The same type of effect can be achieved by rubbing a candle over the surface – candle wax behaves in the same way as beeswax.

need

* ❋ emulsion paint in two colours
* ❋ beeswax furniture polish
* ❋ soft cloth
* ❋ paintbrush
* ❋ wire wool

1 Apply a good base coat to the object to be treated with the resist effect. Two or more coats may be necessary to ensure even coverage. Allow each coat to dry before applying the next, then allow the base to dry completely before continuing.

2 When the paint is dry, rub beeswax furniture polish over the surface of the object with a soft cloth. Do this in a random fashion leaving gaps here and there. Apply a contrasting top coat of paint with a paintbrush. Then set the object aside and allow it to dry completely.

3 When the top coat of paint is completely dry, rub vigorously at the surface with a small pad of wire wool. The wax acts as a resist, so the top coat will not adhere to it and can easily be removed to reveal the contrasting colour underneath.

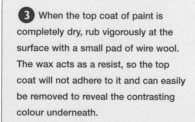

❋ HOW CAN YOU RESIST THIS EFFECT?

WOOD GRAINING

Two paint colours and a clever little graining tool are all you need to achieve this two-tone wood-grain effect. The tool has a plastic handle with a curved pad at one end. The pad has moulded ridges, which, when dragged through the wet top coat of paint, make the wood grain – very clever. You can be a real cheat with this tool by giving a flat MDF surface, or even melamine, a fake wood-grain effect. Try it in funky colours too.

LOOKS LIKE THE REAL THING!

1 Apply one or two layers of the base coat to the surface to be decorated, then allow it to dry. When it is completely dry, apply a thick coat of a contrasting colour on top of the base coat. If treating a large area, add some scumble glaze to the top coat so that it will not dry too quickly.

2 Place the wood-graining tool on the wet top coat and drag the moulded face through the paint, rocking occasionally as you work your way downwards. The rocking action produces the lovely knot effect. Work in vertical rows like this until the surface is complete. Try to ensure that the 'knots' occur in different positions along the surface of the object to be treated.

3 Use bright contrasting colours for a fun look, or more muted natural tones for a real wood effect.

❊ GRAIN WITHOUT PAIN!

need
* two colours of emulsion paint
* wood-graining tool
* paintbrush

CRACKLE GLAZE

This paint effect is designed to emulate the cracked appearance of really old paint. You can buy crackle-glazing kits from good craft shops. The kits contain everything you'll need, including applicator brush and full instructions. The kit I used contained a metallic gold-coloured base coat, the crackle varnish and an ivory-coloured top coat. The products are water-soluble, so the brushes are easy to clean afterwards. The cracked effect is caused by the drying action of the crackle varnish, which works against the top coat, causing the surface of the paint to separate and reveal the base coat. The effect taking place can be seen slowly as the paint dries, but you can use a hair drier to speed it up.

WATCH THE PAINT CHANGE BEFORE YOUR EYES!

need
* crackle-glaze kit
* object to decorate
* hair drier (optional)

1 Using the applicator brush provided, apply a coat of the base coat to the surface to be decorated. Allow the paint to dry completely before continuing with the next stage of the process.

2 Using a clean brush, apply a coat of the special crackle varnish, taking care to brush in the same direction as the base coat.

3 When the crackle-glaze layer is dry, you can apply the coloured top coat. As before, brush in the same direction, and this time take care not to brush over the same area more than once because this can spoil the effect. Set the object aside to dry. The crackle effect will probably begin quite quickly, but the speed of the process depends on the air temperature.

4 Gradually you will see crackles appear all over the surface. If you're in a hurry, you can speed up the process by using a hair drier on medium heat.

* MY PICTURE FRAME IS THE ONLY THING THAT'S SUFFERING FROM PREMATURE AGEING, THANK YOU!

And finally...

LIMING WOOD 100

A subtle treatment for bare or stained wood. The liming wax enhances the grain, lightens the wood and gives a soft, pale finish.

Number 100, and you didn't need a man (for DIY that is)!

need
* stiff wire brush
* liming wax
* wire wool
* soft lint-free cloth
* clear finishing wax

* THAT'S ALL, FOLKS!

1 Make sure that the object to be limed is dry, clean and free of any grease. To begin you must raise the grain of the wood. To do this, brush vigorously with a stiff wire brush and water in the direction of the grain. The moisture and the action of the brush will raise the grain. Set aside in a warm place to dry naturally.

2 Put a little liming wax on a small pad of wire wool. Work on small areas at a time, using small circular motions. This rubbing action works the wax deep into the raised grain of the wood. Cover the entire surface in this way. Set aside for a few minutes to allow the wax to dry.

3 Use the lint-free cloth to rub the surface with clear finishing wax. This removes any excess liming wax and protects the surface. Use a fresh piece of cloth when it becomes clogged. Finally, buff the surface of the object to a smooth sheen, using a pad made from a piece of soft cloth.

Just when you thought it was all over, here are some extra faking ideas

OTHER SUGGESTIONS

Now you know the rudiments of 'faking it' you will find that it's quite amazing what you can do with a coat of paint and a little imagination. I've said before that if a thing is worth doing, then it's worth overdoing – well, in the case of paint effects, less is more! Sometimes a paint effect works better when combined with a plain expanse of colour as a beautiful contrast rather than complete overkill (unless that's what you want of course!). Try an effect on one wall first to see how you like it, then you can always continue with the rest of the room if you really like the look.

Furniture is a different matter, but a room can most definitely stand over-the-top items, or a focal point of sorts, so this gives a bit of scope for letting go a little. If all else fails, you can always paint over it or dilute it a little – after all it is only paint! We've just touched on a few basic paint effects, but there are so many other great ideas that you could try if you're feeling in the mood for a little bit of experimentation.

THIS KITCHEN SHOWS A COLOURFUL WALL MURAL THAT CONTRASTS WELL WITH THE PLAIN WOOD UNITS.

YOU CAN FAKE IT EVERY DAY WITH A BLACKBOARD WALL! JUST CHOOSE YOUR DESIGN AND THEN RUB IT OUT WHEN YOU FANCY A CHANGE OF DECOR!

*SORRY, I'M TOO TIRED, I'VE BEEN BUSY FAKING IT ALL DAY!

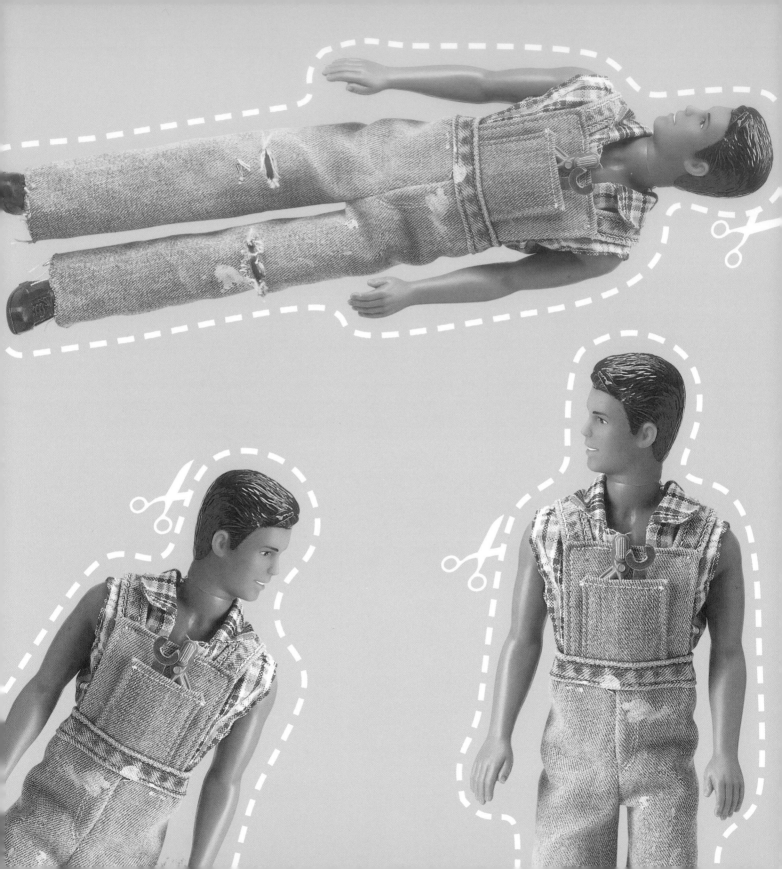

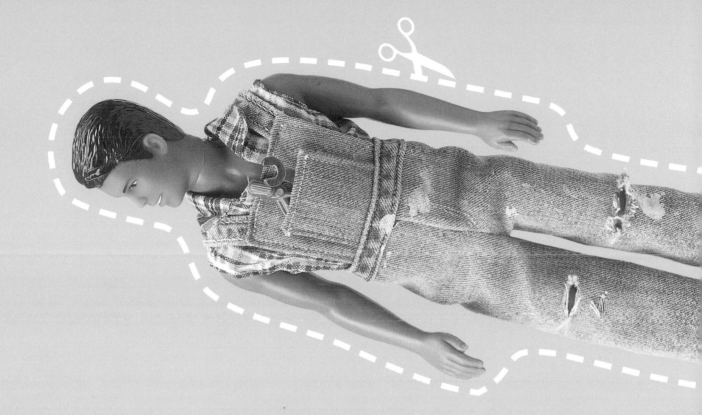

NOT DOING IT

When a man can be useful . . . Guidelines on calling in the experts, how to talk their language and tips on negotiating a fair price for the job.

Here he comes

I guess there really are times when a girl has to do what a girl has to do.

WELL, NOW that you know how to tackle 100 things that you don't need a man for, it's time to look at a few things that do require the attention of the professionals. Although DIY can be an enjoyable activity, you don't want to spend the rest of your life doing it. Sometimes it's best to bite the bullet and make that call to the man who can. However, there are a few things to consider first. Spending a little time and effort to ensure that the man you call in is a true professional really will pay off. Remember, a good reliable workman is worth his weight in gold, so choose carefully.

✳ HMMM... TEN CUPS OF TEA, FOUR PLATES OF BISCUITS, AND THAT PIPE STILL ISN'T FIXED...

WHEN TO CALL A MAN IN

When to call in the workman? Here are a few basic guidelines to help you decide the best course of action. The first few things involve tasks that are dangerous, and then there are those tasks that you may find you have insufficient strength to contend with. Another point to remember is that if the job in question is fairly substantial, you need to get it done before you do any serious cosmetic decoration.

The novice DIY enthusiast should avoid anything that involves serious electrical work, for example fitting new sockets, re-routing cables or rewiring. Older properties may have very old wiring systems that need to be completely replaced. This is not impossible of course, but as far as electricity is concerned, you really need to know what you're doing. For your own safety and that of others, this is when you need to call in the professionals. Gas appliances, such as cookers or heaters, should always be repaired or installed by a registered or approved technician. Like electricity, gas is very dangerous and so under no circumstances should you take any chances with these appliances.

Major plumbing work requires the skill and strength of a professional. Hauling baths and other bathroom furniture around does not appeal to me in the slightest. Neither does laying miles of pipework under the floor or clearing blocked soil stacks or anything to do with major drainage systems.

Dampness and any subsequent effects to timber, structural damage due to problems with the foundations or damage to the exterior or roof can be very serious and costly. Do not attempt to fix these yourself, unless the problem is a simple blockage such as a clog in a gutter (*see page 48*).

IT'S LIFE... BUT NOT AS WE KNOW IT

JUST SO YOU KNOW EXACTLY WHAT YOU'RE DEALING WITH!

THE WORKMAN

Workmen are a completely different species, so a girl needs to wise up to their habits. The workman has a certain characteristic that precedes talking about money: a slight pursing of the lips, a sharp intake of breath usually combined with a scratch or shake of of the head. Then they say: 'It's gonna cost you!' The level of drama involved in this procedure will indicate to you the level of 'cowboy' in the man. This, in combination with baffling terminology, should suggest that you thank the man for his time, then cross him off your list.

ACCESSORIES

- ☞ Cup of tea with four sugars
- ☞ Newspaper
- ☞ Muddy shoes
- ☞ Multitude of facial expressions
- ☞ Inept assistant (optional)
- ☞ Pencil behind ear
- ☞ Degree in 'pocket philosophy', not unlike a taxi driver

Lumberjack shirt

Usually spattered with plaster or paint for that authentic 'I've been working really hard' look.

Tea

This man exists in a state of semi-consciousness. But load on the sugar – he needs plenty of calories to do his business.

❋ WHAT HAVE I LET MYSELF IN FOR?

Tools

He'll have a huge bag of tools, but is he going to use any? And watch out for dirty boots on your new carpet too.

Deciding what you can and can't do

MAJOR CONSTRUCTION WORK

BIG JOBS

It is perfectly feasible to remove things such as stud partition walls yourself (and indeed I have done so, with the aid of a friend and a sledgehammer!). However, anything more substantial should first be checked by someone who knows, then handled by someone who can. What is a partition wall and what is not? In general, if you tap a wall and it sounds hollow, then it's likely to be hollow and therefore made of plasterboard on a timber frame (the vertical supports are called studs). If it sounds solid, don't hit it with anything until you assess whether it is crucial to the construction of

* I COULD LAY IT MYSELF, BUT I MIGHT BREAK A NAIL...

the building – you may be friendly with your upstairs neighbours, but you don't want them dropping through the ceiling into the living room.

Do not attempt to remove or alter a supporting or load-bearing wall. It is there, as the name suggests, to support other parts of the building. If you desperately want that open-plan look, or bigger windows or patio doors where there were none, get a professional to do it for you. Doorways and windows have horizontal supports across the top of the opening. If you knock this away without installing another means of temporary support, you really are in trouble. Don't risk it.

Flooring is another area that could require some heavy-duty treatment. Replacing suspended floors and pouring a concrete floor are both heavy jobs. Any large amount of concrete requires a mechanical mixer – lots of shovelling and sweating. To avoid injuring yourself, the sensible option would be to call in the professionals.

NOTE
Remember, if you attempt to fix any appliances and get into problems, it may affect your insurance cover. Manufacturers often ask for payment to repair 'self-inflicted' breakages.

Identify the problem and learn the language

WHAT TO ASK FOR AND HOW TO ASK FOR IT

Now that you've decided when to call in the professionals, you need to know exactly what to ask for. Do your best to familiarize yourself with the problem before making the call. Make a few notes before ringing so that when you have the workman on the telephone you can give him enough information to help the process along. 'My, er, what is it called now, is, um, a bit broken, um, er' is not going to help him work out what the problem is.

Learning the proper names for fixtures and parts, as well as a few technical terms, is a good idea because you will come across as a person who can't be fooled. Even if you can be fooled, don't let him know. The professional tradesman has the advantage of many years of experience and a wealth of vocabulary related to his particular trade that will completely baffle you. A great many of these words are abbreviated terms familiar only to those who have access to the sacred worlds of plumbing and electricity. Try not to look confused – that's just like a red rag to a bull if the person you're dealing with has a bit of the cowboy in him. Remember, knowledge is power!

DO YOUR HOMEWORK

Reputable tradesmen of course wouldn't take advantage like this, but the first rule is: don't trust anyone – yet! Try to get as much information as possible about the job you intend to have done. Go to the library and look for a book on the subject, or have a look around your local DIY stores for free leaflets and information.

You could also go to a few builders' merchants or trade supply shops. I have found that most of the time the staff will be happy to give advice if asked politely. Also, glean as much information as you can from anyone who has had work of a similar nature done. They will already be familiar with the workman's language and will give you a crash course in communication.

TAKE YOUR TIME

Always remember: do not be pressurized into anything. If you are completely confused and full of doubt, then go away and take time to think – taking time out for a day or two won't make any difference. If the workman isn't happy with you doing this, maybe he's not the one you should choose anyway.

BECOME FLUENT IN WORKMAN-SPEAK!

TALK THE TALK

☛ *'It'll take about two weeks.'* Translation: *'It will take about two months.'*

☛ *'I'm just taking a short break.'* Translation: *'I definitely won't be back until tomorrow.'*

☛ *'I'm just going out to get some parts.'* Translation: *'See you in two weeks' time.'*

☛ *'It's gonna cost you.'* Translation: *'I probably can quote you any price because you look like you have no idea what I'm talking about.'*

＊ *SOMETIMES TALKING JUST ISN'T ENOUGH!*

This time you can't rely on your little black book

HOW TO FIND THE MAN

You need the services of a man who can – how do you go about it? The first place you should start is your personal phone book. Ask all your friends who have had any work done to give their opinion about the service they got. You'll find that everyone has their own particular tale of 'workman woe' to relate, but they will be able to point you in the direction of the trustworthy and reliable ones too. Personal recommendation is very valuable, so use it. Many workmen build up good (and bad) professional reputations in

✳ MMMM... BRAD'S GOOD AT WOODWORK AND FRANK IS EXCELLENT WITH HIS HANDS, BUT WHAT I REALLY NEED IS A PLUMBER.

this way, and are usually delighted to get your business through a recommendation from a satisfied customer.

DETECTIVE WORK

No one you know got on the DIY rollercoaster? Look at the phone books for your local area. Workmen usually like to stay within a certain radius of their base, so it's best to choose one near to where you live. Most professional tradespeople are regulated and registered by bodies related to their trade; this is usually indicated by letters after their name or business name. Avoid those who do not state this in their advertisement, or if they do, check and then cross them off the list if they are not officially recognized by the body concerned. Establish a shortlist of maybe six names. Give them all a call, tell them what you need and when; good tradesmen are always busy, so the time schedule will be important for them. Also, if they're not busy you may want to ask yourself why?

GETTING A QUOTE

Under no circumstances agree to anything over the telephone. A contractor should come to your house and make an inspection first and then you'll get your price. Always ask for a written quotation, and ask for a few quotes from different companies. Most will do this free of charge as an incentive to raise new

CHECKLIST

☞ Ask your friends or neighbours for recommendations.

☞ Make a shortlist of companies and get written estimates from them.

☞ Take your time to consider an estimate, don't let anyone pressure you into a decision.

☞ Always get a final estimate with a full specification in writing of the work to be done.

☞ Pay promptly if you are satisfied with the work.

business. A good workman will not try to bamboozle you out of your hard-earned cash; his reputation relies on giving good service. Remember that you are as valuable to him as he is to you.

Always ask for a quote in writing together with a full breakdown of the work involved if possible. When you've got your estimates, sit down and consider the options – the cheapest is not necessarily the best. Does the job include any preparation work? Can you cut costs by doing this yourself, or will you have to do this yourself anyway? Will they be sub-contracting the work? Is the company insured? Will they guarantee the work for a certain period of time? It's worth asking a lot of questions before the deal is settled.

Last but by no means least: always, always, always put the final agreement in writing. This will save a great deal of confusion and frustration in the end. If the contractor you decide to use is a small one, he may not want to issue a contract, but it may be worth making a note of any discussions you have with him regarding work, schedule and part or full payments, so consider writing a few letters of confirmation just to avoid any confusion in the process.

Another point to bear in mind is that if you are issued with a contract or guarantee, make sure that it is worth the paper it is written on – it may contain a lot of technical jargon that in effect doesn't mean much. If in doubt, get a second or third opinion.

HOW MUCH TO PAY HIM

No one likes to overpay for a job, but it is worth paying a reasonable rate for a proper, professional job. Deciding what is a reasonable rate can be the difficult part. Again, ask people who have had similar work carried out, examine your quotes carefully, get second opinions. By taking a little time to collate all this information you will probably be getting a good idea of the average cost you should expect.

If the whole project has escalated and the finances are looking a little undernourished, it may be worth putting the job on hold for a while until you can afford a little more outlay – don't fall into the trap of going for the cheapest option if it means an inferior job. If a job's worth doing, then it's worth paying for it to be done well.

A good tradesman will always be open to a certain amount of negotiation (but don't try to rip him off – he can easily do the same to you). Always be honest and fight fair! Once all the details are agreed upon, let the work begin. Hopefully this experience will be a relatively painless one and you will be delighted with the result – if so, pay promptly. Small businesses or sole practitioners rely on adequate cash flow to survive, so don't make it hard for them. If you find a reliable and efficient professional, it's well worth keeping in his good graces. You never know when you may need something else done in the future.

PROBLEMS

If you do experience problems, say so. Make sure that you and the tradesman know exactly what's happening – keep in touch with each other and get progress reports from him regularly. Let him know if mistakes have been made or if you've changed your mind about any aspect of the project, no matter how small. And if you are so unfortunate as to have a disaster on your hands, make an official complaint to the regulating body with which the workman is registered.

Good workmen are worth their weight in gold.

❋ STRAP YOUR NEGOTIATING GEAR ON AND GET TO WORK!

TIP
When you've made your choice and the work begins, make sure that you have a copious supply of tea and biscuits – it's the workman's staple diet!

House record

Keep this updated regularly — that's what it's there for.

THIS IS A CHECKLIST of reminders and information that will be handy both in emergencies and for regular maintenance procedures. If you don't know where your service panels for the utilities are, then find out and make a note of them on this page now, along with dates of annual services. Keep a record of any problems that have occurred, when they were fixed and who fixed them. It's a really good idea to familiarize yourself with the location of all the important 'controls' of your home when you move in. Keep your notes in a safe place together with any others so that you can pass them on to the new owners if you move out.

WHERE?

Essential services consumer boxes

GAS MAINS ON/OFF	WATER MAINS ON/OFF INSIDE AND OUTSIDE	ELECTRICITY MAINS ON/OFF	CENTRAL HEATING CONTROLS	WATER TANKS
		(In your consumer box make sure you know which fuse governs which circuit, just in case you need to isolate it.)		

WHEN?

Make a note of annual service dates and renewals of service contracts or other important things to remember.

BOILER SERVICE	INSURANCE FOR CONTENTS AND BUILDING	SMOKE ALARM BATTERY CHANGE	SEASONAL CHECK FOR ROOFING, GUTTERING, DRAIN BLOCKAGES, ETC.	OTHER

WHO?

Keep records and guarantees of any work that has taken place: what it was, when it was done and who did it. Maybe you won't need this information, but these guarantees are usually transferable and may be of use to the future owner of your home, should you sell.

DAMP PROOFING	REPLACEMENT WINDOWS	ROOF REPAIRS	BUILDING WORK	NEW FLOORING MATERIALS

A girl can always use a little guidance

TEMPLATES

The templates and guides on the next four pages are for the DIY projects that are featured throughout the book.

You don't have to use these designs; be as wild or as subtle as you want!

Daisy stencil, pages 158—60

Mirror splashback, pages 108—109

Bookcase door, pages 114—15

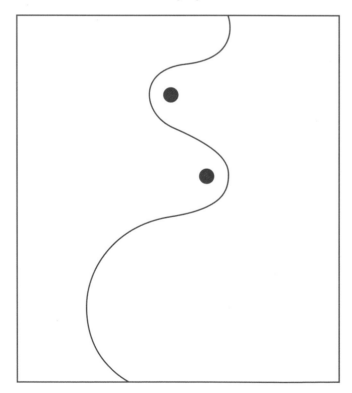

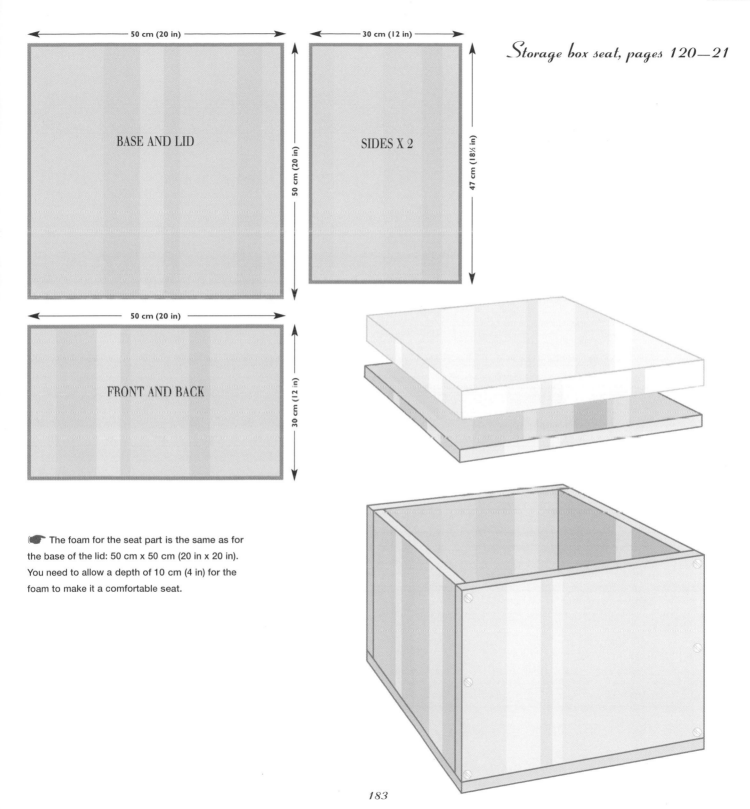

← 50 cm (20 in) →

BASE AND LID

50 cm (20 in)

← 30 cm (12 in) →

SIDES X 2

47 cm (18½ in)

Storage box seat, pages 120—21

← 50 cm (20 in) →

FRONT AND BACK

30 cm (12 in)

☛ The foam for the seat part is the same as for the base of the lid: 50 cm x 50 cm (20 in x 20 in). You need to allow a depth of 10 cm (4 in) for the foam to make it a comfortable seat.

183

Underbed storage project, pages 124—25

☞ You could paint the cut-to-size MDF boards before you assemble the box. Just retouch any scrapes afterwards!

BASE

90 cm (36 in)

61 cm (24 in)

LID

87 cm (34¼ in)

58 cm (22¼ in)

LONG SIDE X 2

90 cm (36 in)

20 cm (8 in)

SHORT SIDE X 2

58 cm (22¼ in)

20 cm (8 in)

Bookshelf, pages 94—95

☞ Not as complicated as it looks!
This shows you how it all slots together.
Slot together shelves 1–5 and then screw
on sides, base and top (6–9).

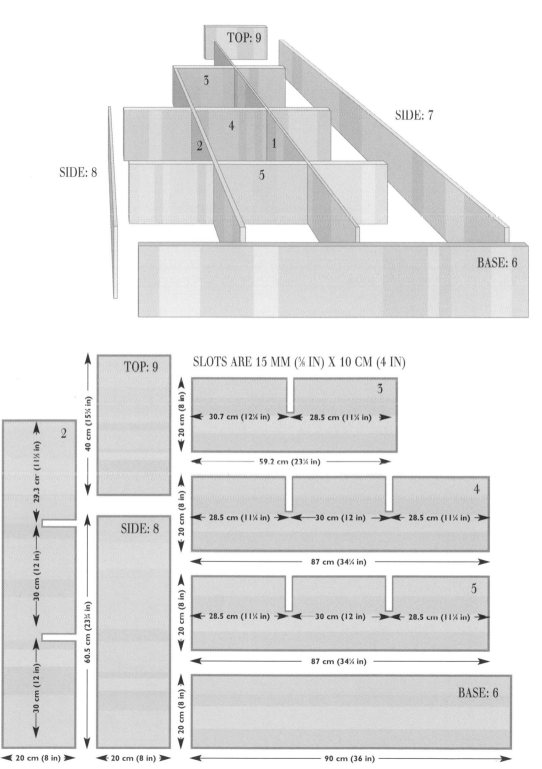

SIDE: 7

SIDE: 8

TOP: 9

3

4

2

1

5

BASE: 6

SIDE: 7

120 cm (47¼ in)

20 cm (8 in)

1

120 cm (47¼ in)

30 cm (12 in)

30 cm (12 in)

30 cm (12 in)

30 cm (12 in)

20 cm (8 in)

2

89.3 cm (35½ in)

29.3 cm (11½ in)

30 cm (12 in)

30 cm (12 in)

20 cm (8 in)

TOP: 9

40 cm (15¾ in)

SIDE: 8

60.5 cm (23¾ in)

20 cm (8 in)

SLOTS ARE 15 MM (⅝ IN) X 10 CM (4 IN)

20 cm (8 in)

3

30.7 cm (12⅛ in)   28.5 cm (11¼ in)

59.2 cm (23¼ in)

20 cm (8 in)

4

28.5 cm (11¼ in)   30 cm (12 in)   28.5 cm (11¼ in)

87 cm (34¼ in)

20 cm (8 in)

5

28.5 cm (11¼ in)   30 cm (12 in)   28.5 cm (11¼ in)

87 cm (34¼ in)

20 cm (8 in)

BASE: 6

90 cm (36 in)

GLOSSARY OF TERMS HUH?

ALCOVE Floor-to-ceiling recess in walls, usually either side of a chimney breast.

ARCHITRAVE Strips of timber moulding fixed around door or window frames as a decorative means to hide the join between the frame and the surrounding wall.

AUGER Long flexible rod used to clear blockages from water and waste pipes.

BALLCOCK Float-operated valve installed in water tank or cistern to regulate water flow and refill when emptied.

BASE COAT Applied over primer and before the top coat of a paint system.

BATTEN Narrow strip of wood.

BEADING Narrow strip of moulded timber used as a decorative edge or frame.

BEVEL A slanting or sloping edge.

BIT Metal attachment for hand or power drill to facilitate drilling of holes of various diameter.

BLOWN When a layer of plaster or render comes away from a wall.

BULLNOSE Tread at the base of a flight of steps that has a curved edge or nose.

BUTT Join between two pieces of wood. The pieces are cut square and one is butted to its neighbour.

CASEMENT WINDOW Window with one fixed pane and two hinged panes.

CASTOR Wheel attached to the underside of furniture to render the piece easily mobile.

CAVITY WALL Wall made up of two layers of masonry or plasterboard with an air space in between.

CISTERN Water storage container used in plumbing systems and flushing mechanisms for WCs.

CLEARANCE HOLE Guide hole drilled through a piece of wood before countersink holes are made or screws inserted.

CONSUMER UNIT The control panel of your home's electrical supply.

COVING A plaster, polystyrene or sometimes wooden moulding used to cover the join between the wall and ceiling.

CROSS-HEAD Term for screw with cross-shaped indentation in the head, for use with cross-head (Phillips) screwdrivers.

DADO This separates the lower part of an internal wall from the upper part by a rail.

DOWEL A small cylindrical peg used to form a secure joint between two pieces of wood.

FINIAL Decorative end attached to a curtain pole.

FLAP VALVE Circular rubber valve found in the siphon inside a toilet.

FLOAT see BALLCOCK.

FLOGGER BRUSH Brush with long floppy bristles used for paint effects.

✳ SO THAT'S WHAT A FLOGGER BRUSH IS FOR!

FLUSH When two surfaces are perfectly level with each other.

FUSE A protective device used in some electrical systems to prevent current overload.

GLAZING SPRIGS Small nails used to secure a glazed pane into a window frame.

GRAIN The direction or pattern of wood fibres.

GRAINING Paint effect to simulate wood grain.

GRIPPER ROD Narrow timber strip with rows of small spikes used to secure carpets.

GROUT Used to fill and seal the gaps between ceramic tiles after they have been fixed to the walls. Used also for mosaic tiling projects.

GULLY Trap into which rainwater and waste pipes discharge before emptying into the drain.

HAMMERED FINISH Metallic spray paint designed to imitate a hammered-metal finish.

JOIST Horizontal or metal supporting beam for walls, ceilings or floors.

JUMPER Lower part of tap headgear to which the rubber washer is attached.

KEY Term applied to a surface that has been sanded ready for painting or gluing.

MASKING TAPE Low-tack adhesive tape used to mask off areas when painting.

MDF Medium-density fibreboard made from compressed wood fibres with two smooth sides. MDF has similar properties to wood and is available in various thicknesses.

MELAMINE Chipboard that is laminated on one or both sides.

MISCOAT Coat of watered-down emulsion paint applied to bare walls to accentuate holes, cracks and areas requiring filling.

MITRE Joint between two pieces of wood where the ends of each piece have been cut to the same angle.

MORTISE Rectangular recess cut into a piece of timber or a door to receive a lock or another piece of timber cut to form a tenon.

NAP The direction in which fabric fibres lie.

'O' RING Rubber seal found in a tap.

PAD SAW Saw with narrow blade for cutting small holes in wood.

PARE Using a chisel to remove slivers of wood.

PILLAR TAP Name for old-style tap.

PILOT HOLE Preliminary hole drilled into sheet material to facilitate easy insertion of a jigsaw blade or small-diameter hole used as a guide for a screw.

PLASTERBOARD Board made of layers of fibreboard or paper adhered to a gypsum plaster core, used instead of plaster or wood panels to form walls.

PLUMB LINE The true vertical line, or the name for the weighted string used to find the vertical.

POST TERMINAL Screw fitting that holds electrical wires in a plug.

PRIMER First coat of paint system.

REBATE Rectangular recess found along the edge of a frame or moulded workpiece.

RESIST EFFECT Paint effect where a wax substance is applied over a base coat; the top coat is then rubbed away to reveal the base because the paint will not adhere to the wax.

RISER Vertical portion of a step.

SASH Term applied to the sliding parts of a double-hung window.

S-BEND Trap found under a sink or basin, designed to catch debris before it enters the wastepipe to avoid blockages.

SCREED Layer of cement used to level a floor.

SCRIBE Using a spiked batten to trace the profile of a surface along the edge of sheet material.

SHIM Wedge-shaped pieces of wood used as supports.

SHROUDED-HEAD TAP Term for modern-style tap.

SLOT-HEAD Screws with a single slot-shaped indentation across the head.

SOIL PIPE Large-bore vertical pipe that takes waste sewage to a drain.

STOPCOCK Tap on a water supply system used to turn the flow off or on.

STUD Vertical timber which forms part of a plasterboard partition wall.

STUD PARTITION Wall constructed from wood framework faced with plasterboard.

TAMP Using a cloth or brush to ensure a material is properly adhered to another surface or packed down firmly.

TONGUE AND GROOVE Term applied to wooden panelling that has a tongue down one edge and a groove down the other. The panels slot together.

TOP COAT The final coat of a paint system.

TREAD Horizontal part of a step.

UNDERLAY Foam layer placed over a subfloor before laying woodstrip flooring or a carpet.

WALL PLUG Plastic sleeve to fit a screw. The wall plug is inserted into a pre-drilled hole in a wall. When the screw is inserted, it then expands to form a firm gripping surface for the screw's thread.

WASHER Rubber disk found in a tap.

WIRE WOOL Fine metal fibres gathered together to form a pad used for sanding.

WOODSTRIP A form of wooden flooring available as slotted strips or planks.

FURTHER READING

NOW FOR SOME BEDTIME READING!

1,001 DIY HINTS AND TIPS
(Reader's Digest, 1999)

COLLINS COMPLETE DIY MANUAL
Albert Jackson and David Day (HarperCollins, 1998, revised 2001)

COLLINS PLUMBING AND CENTRAL HEATING
Albert Jackson and David Day (HarperCollins, 1999)

THE COMPLETE BOOK OF PAINT TECHNIQUES
Penny Swift and Janek Szymanowski (New Holland, 1994)

THE COMPLETE DIY MANUAL
Mike Lawrence (Lorenz Books, 1999)

THE COMPLETE GUIDE TO WALLPAPERING
David M. Groff (Creative Homeowner Press, 1999)

COMPLETE PAINT EFFECTS
Sacha Cohen, Maggie Philo (Lorenz Books, 1999)

CUPBOARDS AND DOORS IN A WEEKEND
Deena Beverley (Murdoch Books, 1999)

DECORATIVE PAINT FINISHES
Louise Hennings and Marina Niven (New Holland, 2000)

HANDY ANDY'S HOME WORK
Andy Kane (BBC Publications, 2000)

HOME FRONT: STORAGE
Tessa Shaw (BBC Consumer Publishing, 1998)

THE NEW COMPLETE BOOK OF DECORATIVE PAINT TECHNIQUES
Annie Sloan and Kate Gwynn (Ebury Press, 1999, 3rd edition)

ON THE SHELF
Alan and Gill Bridgewater (New Holland, 2000)

PAINT MAKEOVERS FOR THE HOME
Sacha Cohen (Southwater, 2000)

THE POCKET ENCYCLOPEDIA OF HOME REPAIR
John McGowan and Roger Du Bern (Dorling Kindersley, 1991)

READER'S DIGEST COMPLETE DIY MANUAL
(Reader's Digest, 1998)

THE WALLPAPERING BOOK
Julian Cassell and Peter Parham (Haynes, 1996; reprint 1997)

THE WALL TILING BOOK
Alex Portelli (Haynes, 1996)

THE WEEKEND CARPENTER
Philip Gardner (New Holland, 2000)

THE WHICH? BOOK OF DO-IT-YOURSELF
(Which? Books, 1999)

THE WHICH? WAY TO FIX IT
Mike Lawrence (Which? Books, 1999)

WOODWORKER'S HANDBOOK
Roger Horwood (New Holland, 2000)

❋ I'LL LEND YOU MY GUIDE TO ADVANCED GROUTING IF YOU COULD JUST LEND ME THOSE RED SPIKE HEELS.

USEFUL ADDRESSES

Try your local DIY store first, but if you have trouble locating any products try the following suppliers:

PAINTS AND FINISHES

CROWN PAINTS
Tel: 01254 704951

www.crownpaint.co.uk

DULUX
Tel: 01753 550555

www.dulux.co.uk

HAMMERITE
Tel: 01661 830000

www.hammerite.co.uk

INTERNATIONAL PAINTS
Tel: 01962 717001/002

www.plascom.co.uk

RONSEAL
Tel: 0114 2467171

www.ronseal.co.uk

FLOORING

ALLIED CARPETS
Tel: 01689 895000

www.alliedcarpets.co.uk

Wide range of floor coverings

AMTICO
Tel: 0800 667766

www.amtico.com

Vinyl flooring

CRUCIAL TRADING
Tel: 0800 374429 for brochure

Natural floor coverings

PERGO
Tel: 0800 374771

www.pergo.com

Laminate and floorboards

SLATE WORLD
Tel: 020 8204 3444 or 020 7384 9595

www.slate-world.com

WALLCOVERINGS

COLOROLL
Tel: 0800 056 4878

www.coloroll.co.uk

Wide range of wallpaper

GRAHAM & BROWN
Tel: 0800 3288452

www.grahambrown.com

Contemporary wall textures

LAURA ASHLEY LTD
Tel: 0800 868100

www.laura-ashley.com

Traditional wallpapers, paints and paint effects

TILES

FIRED EARTH
Tel: 01295 814315

www.firedearth.co.uk

Ceramic, terracotta, stone and slate tiles

JUST TILES
Tel: 020 8907 3020

www.justtiles.co.uk

Discount floor and wall tiles

ORIGINAL STYLE TILES
Tel: 01392 474011

www.originalstyle.com

Victorian and other ranges of tiles

SUNDRIES

CURIOUS PEDESTRIAN
Tel: 01453 886482

Wacky toilet seats

JALI
Tel: 01227 831710

For radiator cover kits, shelving, etc.

DIY SUPERSTORES

B&Q
Tel: 0845 309 3099

www.b-and-q.co.uk

FOCUS DO IT ALL
Tel: 0800 436 436

www.focusdoitall.co.uk

HOMEBASE
Tel: 0845 300 1768

www.homebase.co.uk

WICKES
Tel: 0870 608 9001

www.wickes.co.uk

OTHER USEFUL WEBSITES

DIY DOCTOR
www.diydoctor.org.uk

DIY FIXIT
www.diyfixit.co.uk

ONLINE DIY
www.onlinediy.co.uk

Index

Now where did I read about clamps and floggers before?

A

ABRASIVES 24, 132
ADJUSTABLE SPANNER/
WRENCH 11, 19
ALCOVES, SHELVING 86
ALLEN KEYS 19
ANTIQUING 131, 166,
167

B

BAGGING 155
BALLCOCK 41
BATH PANEL 102–103
BATHROOM
bath panel 102–103
blocked sink 49
leaky plug 43
mosaic splashback
108–109, 182
new showerhead 43
tiling shower 107
toilet 40–43
BED, STORAGE 122,
124–25, 184
BLACKBOARD WALL 171
BLEACHING
floorboards 59
BLOCKAGES
drain 48, 174
guttering 48, 181
plug hole 49
toilet 41
BOLT, FITTING 36
BOOKSHELF see
shelves
BORDERS
carpet 77

stencilling 76, 158
wallpaper 135
BOX
seat 120–21, 183
underbed 124–25,
184
BRACKETS 87, 88
BRADAWL 11
BRICKS, IN SHELVING
118
BRUSHES 14, 15, 150,
156, 159
BUTTERFLY TOGGLE 21

C

CARPET
borders 77
choice 60, 61
laying 72–73
stair 74–75
tape 25, 72
tiles 60, 66
CEILINGS 146, 148, 151
CERAMIC TILES
cutters 22, 23, 105,
106
drilling 107
fixing 104–106, 130
floor 62, 63, 68–69
grouting 106
painting 101, 147,
161, 171
spacers 23, 69, 105,
106
tools/equipment
22–23
wall 104–106, 130

CHISELS 11, 22, 23, 94
COLOUR WASHING 131,
152
COMPASS 19
CONCRETE 77, 176
CORK TILES 61
CORNERS
mitres 54, 55, 82
storage area 122,
123
wallpapering 142–43
COVING 83
COWBOYS 175, 177
CRACKLE GLAZE 16
CRACKS, FILLING
138–39
CROWBARS 17
CURTAINS
pelmet 92
poles/tracks 90–91

D

DADO RAIL 82
DAMP 138, 176
DECORATING
equipment 14–15
planning 128–30
preparation 130,
138–39, 148
DECOUPAGE 171
DESK, SELF-ASSEMBLY
85
DISHWASHER 44–45
DISTRESSING 166, 167
DOORS
chain 36
cupboard 89
direction change 98
draughts 30, 31
fish-eye viewer 35
hanging 97
knobs 99, 165
locks 37
painting 149
refrigerator 46

removing 96
for shelves 114–15,
182
shelving over 122
sticking/squeaking
30
DRAGGING 131, 156
DRAINS 48–49, 174, 181
DRAUGHT EXCLUDER
STRIPS 31
DRAUGHTS 30, 31
DRAWERS, RUNNERS 47
DRILLS 12, 80
DRY BRUSHING 156
DUST SHEETS 15

E

ELECTRICITY 28–29, 89,
143, 174, 176, 180
EMULSION PAINT 147
ESTIMATES 178

F

FASTENERS 20–21
FILLING KNIFE 15, 22
FILLERS 15, 25, 138
FISH-EYE VIEWER 35
FLOORBOARDS
decorative ideas
76–77, 161
finishing 58–59
repairing 53
sanding 56–57, 130
FLOORING 52
see also carpet tiles
choice 60–61
parquet 61
professional help 176
skirting boards
54–55
suppliers 189
vinyl 61, 67, 70–71
woodstrip 60, 64–65
FLOWERPOTS,
SHELVING 119

FLUSH, PROBLEMS 41, 42
FURNITURE
flatpack 84–85
painting 165
special effects 166,
171
stripping 164
FUSES 25, 29, 180

G

G-CLAMP 18
GAS 174, 180
GILDING 171
GLASS
bricks 119
broken 32–33
shelving 119
GLOSSARY 186–87
GLOSS PAINT 146, 147
GLUE GUN 13
GLUE SPREADER 17, 23
GRAINING 131, 168
GRIPPER RODS 73, 74
GROUT SHAPER 23

GROUTING 106
GUTTERING 48, 181

H

HAMMERS 11, 16, 85
HANGING RAIL 113
HINGES 47, 96, 97
HOLES, FILLING 138–39
HOOKS 122
HOT AIR GUN 13, 133,
164
HOUSE RECORD 180–81

J

JIGSAW 12, 80

K

KEY TO SYMBOLS 7
KITCHEN
blocked sink 49
cupboards 47, 89, 100,
101, 165
refrigerator 46,
162–63

❋ DOING IT
YOURSELF IS GREAT,
BUT IT AIN'T NOTHING
LIKE THE REAL THING!

CENSORED

splashback 101, 104
storage ideas 122
worktop 100
KNIVES 11

L
LADDERS 18, 122, 145
LIGHT SWITCHES 143
LIMING WOOD 170
LOCKS 35, 37

M
MARBLING 131
MASKING TAPE 15, 25
MELAMINE PAINT 147, 165
METAL/METAL EFFECT
gilding 171
paint 147, 162–63
tiles 63
worktop 100
MIRROR
hanging 81
mosaic splashback 108–109, 182
MITRES 17, 54, 55, 82
MOSAIC 63, 101, 108–109, 171, 182
MURALS 171

N
NAILS 20
NATURAL FIBRES 60, 61, 73

O
OFFICE SPACE 123
OIL-BASED PAINT 146

P
PAINT PADS 14, 150
PAINTING see also
special effects
behind radiator 151
ceilings 146, 151
doors/windows 149
floorboards 59
furniture 165
method 148–49

old wallpaper 129
paint choice 146–47
preparation 129, 138, 148
stripping 133, 164
suppliers 188–89
tiles 101, 147, 161, 171
walls 146, 150
wood 132, 166
PARQUET FLOORING 61
PELMETS 92
PENCILS 10
PICTURE RAILS 81, 83
PICTURES, HANGING 20, 81
PLASTER 120, 138,180
PLASTERBOARD
fixings 21, 81
removing 176
PLASTIC WALL PLUGS 21
PLIERS 10, 19
PLUG
electric 28–29
sink/bath 43
PLUMB LINE 14
PLUMBING 38–45, 89, 174, 180
POWER TOOLS 12–13, 80
PUTTY 32–33

R
RADIATOR
boxing in 116
painting behind 151
shelves 117
RAG ROLLING 154
RAGGING 131, 154
RECYCLING 96, 118
REFRIGERATOR 46, 162–163
RESIST EFFECTS 167
ROLLER BLINDS 93
ROLLERS 14, 15, 150, 157
RUBBER MALLET 16, 85

S
SAFETY 80
drilling 89
equipment 24
paint stripping 133
painting 162
professional help 174, 176
sanding 57, 139
tiling 104, 106
varnishing 58
SANDERS 13, 133
SANDING
floorboards 56–57
method 130, 132–33
SANDPAPER 24, 132
SAWS 11, 12, 17, 23
SCALE DRAWINGS 76
SCISSORS 15, 19
SCREWDRIVERS 10, 13, 80
SCREWS 21
SEAM ROLLER 15
SEALANT 107, 138
SELF-ASSEMBLY
FURNITURE 84–85
SHAVE HOOK 15, 164
SHELVES
bookshelf 94–95,185
building 86–88
with doors 114–15, 182
ideas for 118–19
over radiator 117
wardrobe 112–13
SHOWER
changing head 43
tiling cubicle 107
SINK, BLOCKED 49
SKIRTING BOARDS 54–55
SLIDING BEVEL 16, 87
SMOKE ALARM 34, 181
SOCKET SET 19
SOLVENTS 15
SPECIAL EFFECTS 131, 166
antiquing 131, 166, 167

bagging 155
colour washing 131, 152
crackle glaze 169
decoupage 171
distressing 166
dragging 131, 156
gilding 171
liming wood 170
mosaic 171
murals 171
ragging 131, 154
resist effects 167
roller effects 157
sponging 131, 153
stamping 161
stencilling 76, 158–60
tiles 171
wood graining 131, 168
SPIRIT LEVEL 10
SPLASHBACK 101, 104, 108–109, 182
SPONGING 14, 131, 153
STAIRS
carpet 74, 75
storage area 123
STAMPING 161
STAPLE GUN 18
STENCILLING
floors 76, 161
method 158–60, 182
refrigerator 163
STIPPLING 131
STORAGE
box seat 120–121, 182
shelving 112–119
underbed box 124–25, 184
wasted space 122–23
STRAIGHTEDGE 11
STRING 14
STUD FINDER 17, 89
SUPPLIERS 188–89
SURFORM FILE 17

T
TAP, DRIPPING 38–39
TAPE MEASURE 11
TEMPLATES 71, 76, 105, 182–85
TEXTURED PAINT 147
TILES see also ceramic tiles
carpet 60, 66
cork 61
flooring 62–63, 66–69
parquet flooring 61
suppliers 189
vinyl 61, 67
TOILET 40–43
TONGUE AND GROOVE 64–65, 102–103
TOOLBOX 25
TOOLS 10–25, 80
TRADESMEN see workmen
TROMPE-L'OEIL 77
TROWEL 23
TRY SQUARE 11

V
VARNISH, FLOOR-BOARDS 58
VINYL
flooring 61, 70–71
painted 77
tiles 61, 67
wallpaper 135

W
WALLPAPER
air bubbles 140
air vents 144
choosing 134–35
corners 142–43
equipment 15
hanging 140–45
light switches 143
old 129
paste 15
patterned 144
preparation 138–40

steamer 13, 137
stripping 136–37
suppliers 189
window recesses 145
WALLS
blackboard 171
cracks/holes 138–39
mouldings 81, 82–83
murals 171
removing 176
skirting boards 54–55
WARDROBE 112–13
WASHERS 38–39
WASHING MACHINE 44–45
WATER-BASED PAINT 146, 147
WEBSITES 189
WINDOWS
broken 32–33
locks 35
painting 149
papering recesses 145
WIRE STRIPPER 17
WOOD
graining 131, 168
liming 170
painting 132
stain 59
WOODSTRIP FLOORING 60, 64–65
WORKBENCH 16
WORKMEN
calling in 130, 174–77
finding 178–79
house record 181
paying 179

Acknowledgements

AUTHOR'S ACKNOWLEDGEMENTS

Thanks to Mam and Dad for their constant support and for being a fantastic pair of folks, to Marie and Meghan, and to all the wonderful girls (and boys!) in my life – thank you, I am very, very lucky. Thanks also to Mike, Vaso and all my family, and to Adrian, Daisy and William (and Aphrodite!). Linda, thanks for the new strategies and for helping me find a window when it was dark. And finally, thanks to everyone at Ivy Press. It's been hard work making this book, but I've had a lot of fun too!

☞ The author and publisher would like to thank the following for their assistance and for the loan of props:

Black & Decker (power tools)

Bryden's DIY, Hove

Chubb (security products)

Curious Pedestrian (toilet seat)

Faithfull (tool box and belt)

Harris (decorating equipment and hand tools)

Heuga (carpet tiles)

International (floor paint, melamine paint)

Laura Ashley (wallpaper)

Paint Magic, Hove

Plasticote (spray paints)

Stanley (hand tools)

PICTURE CREDITS

Abode 52l, 100bl;

Corbis /Jacqui Hurst 118bl /Elizabeth Whiting & Associates 122br;

Dulux (www.dulux.co.uk) 128cr, 146tc;

Graham & Brown (www.grahambrown.com) 134tc;

Harlequin (www.harlequin.uk.com) 134cr;

House & Interiors 119tc;

Ikea (www.ikea.com) 52r, 122cr;

Original Style (www.orginalstyle.com) 23cl;

Mark Wilkinson Furniture (www.mwf.com) 89bl, 171bl;

Verne 171cr.

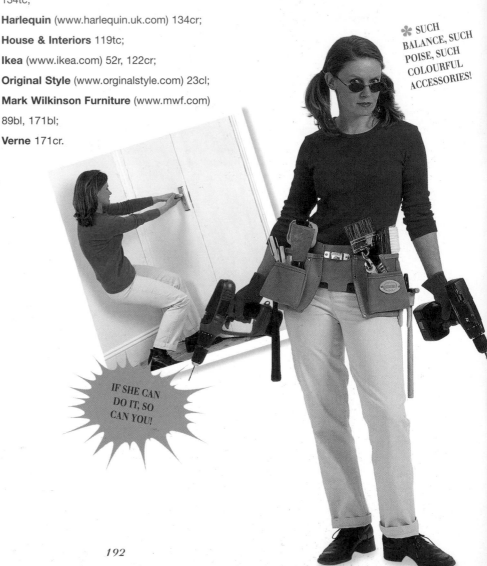

✱ SUCH BALANCE, SUCH POISE, SUCH COLOURFUL ACCESSORIES!

IF SHE CAN DO IT, SO CAN YOU!